THE ANIMALS IN THEIR ELEMENTS

Cynthia Flood

Talonbooks Vancouver 1987

published with the assistance of the Canada Council

Talonbooks
201 / 1019 East Cordova Street
Vancouver
British Columbia V6A 1M8
Canada

Typeset in Goudy Oldstyle by Pièce de Résistance Ltée;
printed and bound in Canada by Hignell Printing Ltd.

First printing: October 1987

Some of these stories have appeared, in slightly different form, in the periodicals *Atlantis*, *Fireweed*, *Journal of Canadian Fiction*, *Matrix*, *Prism*, *Queen's Quarterly*, *Room of One's Own*, and *Wascana Review*; and in the anthologies *Common Ground*, *NEW: West Coast Fiction*, and *Vancouver Short Stories*.

Canadian Cataloguing in Publication Data

Flood, Cynthia, 1940-
The animals in their elements

ISBN 0-88922-249-5

I. Title.
PS8561.L65A76 1987 C813'.54 C87-091450-2
PR9199.3.F56A76 1987

For my daughters, Isabel and Margaret

Contents

Summer's Lease

My grandmother died in May, and that summer my mother and my brother and I went east for two months to the old house in Toronto, so my mother could Clear it out. Sometimes she called it that, and sometimes Emptying that bloody museum, and sometimes Settling up Grandmother's affairs, depending on whom she was talking to. I liked Bloody museum the best, though I did not understand why she spoke the words in such a frantic tone to my father and her close friends. I was nine. William was six. I had never been to a real museum but had read about them and seen pictures. William did not know what a museum was, but he got excited when I told him about glass cases with things in them that were hundreds, thousands of years old. Boxes, jewellery, swords, coins. He wanted to know if there were ghosts, and then our mother came into his room—this was shortly before we left Vancouver, and he was in bed recovering from an ear infection—and told me for heaven's sake not to get him upset, she had enough on her hands already.

I remember still how astonished I was as the taxi drove us from Union Station up to the Annex—astonished at the houses. They looked like castles to me, these giant turreted piles of dried-ketchup brick, draped in ivy. Did people really live in these buildings? Our Vancouver house was on one storey, stucco painted white, with floor-to-ceiling windows facing north to the mountains. My mother was silent, looking quickly left and right and left again. Her face had a tight look which I knew meant she was nervous, but I could not imagine why. William was opening and shutting the ashtray fitted into the back of the front seat. Normally she would have told him to stop that, but she did not appear to notice.

"Oh," she said suddenly, "it's gone."

We were passing a small apartment block.

"My school. The school I went to when I was a little girl used to be right there. I thought we would see it when we came round this corner."

"Not been here in a while, eh?" said the taxi-driver.

"No, not for many years."

"You have so," I said, "you came to Toronto for Grandmother's funeral."

"But I didn't go to the house, Megan. I stayed at a hotel. Oh, here's the park."

The park took up an entire small block; sidewalks enclosed it. Paths ran from each corner to a large circular flower bed in the middle. I could see swings and a water fountain and some organized shrubbery. Nothing special, nothing like our park off Wall Street in Vancouver with its prospect of the harbour and the bridges and the mountains, but my mother's eyes were very wide open.

The taxi stopped. I was disappointed. Not a castle. It was brick, though; a medium-sized two-storey house with a deep front porch. There were fat wooden pillars on the porch, painted dark green, as were the door and the window frames and the shutters. There was a small front garden. The houses on either side were identical, and so were the houses past them.

"It seems smaller," said my mother. "I suppose that's always the way."

The air in the house was dreadfully stale. Underneath the brownish smell were lemon oil, floor wax, Javex, and an aromatic fragrance I found later was pot-pourri. William went rather pale.

There were four bedrooms in the house. I got to sleep in the one that had been my mother's, and William went into what had been our grandfather's study.

"I don't know what the hell he ever studied," said my mother, looking at the bookcase which contained nothing but *National Geographics*, packed so tight I had to help William get the first one or two out.

"Then what did he use it for?" I asked.

My mother snorted. "To keep his sanity I should think. Here William, this bottom drawer's empty, you can put your things in here."

"What do you mean, his sanity?"

"Not now, Megan, for goodness' sake. Did you look at the books in my room? There's a lot of things there I read when I was your age. Oh dear, this is weird."

"What's weird?" But she was gone, and I went to her, my, room and knelt by the bookcase and found all of L. M. Montgomery. I had never been able to get *Emily Climbs* at the Hastings Library and I went right into it. After a while my knees hurt and I got up and looked out the window. This room and Grandfather's study overlooked the garden at the back of the house. It was not like any garden I knew at home. The only flowers were roses. They were ranged like a border, in full bloom and multicoloured, around the neat rectangle of lawn. That was all. I counted thirteen different colours among the roses. I went to tell my mother and found her unpacking in what was clearly the guest room.

"Why aren't you going in the big room next door?" I asked, glancing in at the tall dark four-poster bed with its long fringed white spread. My mother did not answer. She was sitting on the single bed in the guest room looking at a small glass globe in her hands. It was one of those paperweights that you shake to make the snow fall. There was a tiny ballerina tiptoe in the blizzard.

"Silly thing," she said with pleasure in her voice. "Your grandfather gave me that for my birthday once. I guess I was eight or so. Why aren't I sleeping in there? Because I'm sleeping here, that's why."

"The garden," I said, "it's full of roses."

"Your grandfather was a great gardener. He loved flowers. When he was too ill to work in the garden any more he taught your grandmother how to do it. Where's William? We've got to go out and get some dinner, it's too late to shop."

We dug William out of a pile of *National Geographics*—he had dozens of them all over the floor. My mother began to be mad and then said, "Oh well, it doesn't matter I suppose, all that stuff's going out anyway." We walked up to Dupont Street and found a little soda-fountain kind of place, and had hamburgers and French fries, which last we were not allowed at home. It was strange, meeting people occasionally as we walked and not seeing any known faces; and I began to think that we did not know anybody there in Toronto except my mother, and how awful it would be for William and me if we somehow lost her.

In a way that feeling lasted the whole summer. Of course William and I were not alone, for we found children to play with in the park. But we did not make any real friends, and in any case many of them went away to camp or to cottages. Nor was it that our mother did not spend time with us. As I see now, she was very conscientious about ensuring that the summer which was work and much else for her was a real holiday for us. Every few days we went on some excursion, to Centre Island, to the Science Centre, to Casa Loma, to the zoo, to High Park. We saved the trip to the Royal Ontario Museum for last. Almost daily she took us on the subway, which William especially loved. We would choose a stop with a name he liked, and get out there and explore for a while, and then ride back to the Spadina station and walk up to our grandparents' house again. My mother let us stay up late, and watch far more television than we did at home, and have popsicles almost every time we wanted them, and she made us our favourite things to eat so readily it was hardly any fun asking.

But she was abstracted, preoccupied. William especially found it maddening that he would ask her a question and she would not respond, or he would tell her something and find out later that she literally had not heard him. He even cried sometimes in frustration, and then of course my mother got very upset. I think he felt it so much because he had been accustomed to such quantities of attention from her always. He was often sick, ear infections and bronchitis and allergies and skin trouble, and of course it was she who cared for him. I don't mean that my father was indifferent, not at all, he did and was a lot more for us than many fathers as I now recognize; nonetheless he was out of the house all day, five days a week and sometimes six.

One night I was up very late, reading *Rilla of Ingleside* and half in tears over Walter's death, when I heard my mother on the phone in the downstairs hall, talking to my father. I was puzzled, for this was not the routine. We called him every Wednesday and he called us every Sunday, and this was Friday. Her voice was shaky and my stomach went tight.

"But what in hell am I going to do with them? It's a whole bloody boxful of letters, Gerry. Letters she wrote to him when they were engaged, letters he wrote to her. And all of my letters to them, and to her after he died. All of them, can you believe it?"

"I can't bring myself to. I just can't bring myself to read them."

"I can't do that either. It doesn't feel right just to throw them out. Oh I'm sorry darling, I know I'm being inconsistent, but I just can't. I guess I'll just hold on to them over the summer and see."

"Yes, I think it's the house. I can't breathe here any more than I ever could. I feel like I'm in a grave and trying to heave the lid off the coffin."

"Yes, only a few more weeks. That doesn't help."

"I don't mean to be snappy, darling. It's just hard, that's all. I didn't think it would be this hard. I feel as if she's watching me all the time."

"Yes, exactly. And telling me that I'm doing it all wrong and making all the wrong decisions."

"I know. But somehow her being dead doesn't seem to make that much difference. I wish you were here."

"I love you too."

"Oh they're fine," and her voice was all right again, "I think they're enjoying it. It's so strange, Gerry, seeing them playing where I used to."

"No, I'm feeling better now. We'll talk on Sunday as usual. OK, darling, goodbye now." She hung up and said, "Damn my shrew of a tongue," and I heard her go into the living-room and blow her nose and switch on the TV. I turned my light out, feeling that she would come upstairs soon, and she did, and went first into William's room and then came into mine. I lay head into pillow so she could not see my face, knowing from experience that I could not keep my eyelids still. I did not want to talk to her, although I wanted badly to know if Daddy had said that our cat had had her kittens. "Megan?" she said softly. I did not answer. When she had gone I lay awake trying to understand. The dead She must be my grandmother, but why did my mother speak of her so? My mother frightened? Of someone who was dead? I thought of William and ghosts but did not think that was what she meant.

A lot of the work my mother had to do that summer was really quite interesting, and sometimes William and I enjoyed helping her. She went through the house room by room, sorting and deciding what was to be done with what. She did the kitchen first, and the little dark pantry that opened off it. There were piles of kitchen cutlery, forks with bent tines and big dented metal spoons; there were green felt bags full of silver, ingenious pocketed holders with sets of spoons and fish knives, and dinner forks so large William could hardly eat with them.

"Were my grandfather and grandmother very big people?" he asked.

There were cupboards full of great ironstone basins and blue-and-white striped mixing bowls, piles of cream dinner plates with dark green ivy round the borders, dozens of frilly teacups with roses and birds and huntsmen painted on them, well-and-tree serving platters, vegetable dishes with willow-pattern covers, cheese trays and sauce pitchers and gravy boats and preserve dishes all wreathed in flowers.

"She entertained a lot, of course," said my mother, piling up linen dinner napkins. "She used to send me the menus in her letters, and lists of the guests. As if I cared who she had for dinner."

"She wrote letters to you?" asked William.

"Oh my God did she write letters. She. . . ." My mother stopped and would not go on.

At the end of the summer, some of the household furnishings were to go to auction, some to the Salvation Army, some into the garbage. We would ship a few items to Vancouver. My mother bought packets of coloured labels, and William and I went around sticking red ones on auction things and blue on Vancouver things and so on. She said each of us could choose something for ourselves. "Something small." I chose the ballerina paperweight. William would not at first tell us what he had chosen, but finally gave in and showed us that one drawer in Grandfather's desk was filled with empty cigar boxes. "But William," said my mother, "there's nothing special about those, you can get boxes like that in Vancouver any day."

"But these were my grandfather's," said William stubbornly.

"But you never even met him! He was dead before you were born!" She was going to say more but William had that intense look he got—actually, he still gets it—on the verge of one of his upsets, and she said, "Oh all right dear." So William chose a box with a picture of a woman in a long flowing red dress, and put it ceremonially in his suitcase ready to go home.

My mother was now through with the furnishings on the first floor of the house and started in on the papers. There was an old escritoire in the living-room, marked for auction, that was packed tight with them. This was boring, and on afternoons when we were not going out on an expedition William and I took to playing in the back garden. At first there did not seem to be much to do there. We pulled rosebuds off the bushes and they were people and we did long dramas with them, using an orange crate for a house, and after a long time I succeeded in teaching William to stand on his head. We tried to climb the maple tree, but there were no branches low enough to get a foothold on, even when we took a velvet-covered chair out of the dining-room to stand on. Then we found that there was a space between the maple tree and the tall dark green wooden fence at the end of the garden. The ground there was quite bare, and the roots of the maple stuck up out of the earth far enough to be seats. This shaded space, with the dense green leaves above and the tree trunk behind us, became a treasured hiding-place. I sat behind the tree and sent William into the house to check; he looked from both his room and my room and could not see me. So we were safe.

Perhaps the times that we spent there were not really as long as they seem in memory. It seems we spent hours and hours there, and gradually less and less sun sieved through the leaves, and the light that was left was that thick deep gold of late afternoon, and the black earth began to feel chill. Gradually we evolved a play there, a long slow varying play which we did over and over again, that we were very old people, a very old man and a very old woman, the last people left in that country, and when we were gone there would be no one left, no one at all, and our bodies would be very small and dried up—"like mummies," said William, for there had been an illustrated piece on Egypt in one of the *National Geographics*—and so when the leaves fell from the maple tree year after year they would soon cover us, there would only be two small low mounds, and then slowly the earth would build up around us or we would sink into it, lower and lower, and there would be nothing but slender knobby bones gradually flaking and slivering away. And we would lie still, very still, on the earth behind the maple tree, and be so quiet that the song of the cicada sounded like a jackhammer, and close our eyes and feel the wind sifting through the dark green leaves above us. I can hear it yet, that sound of the leaves flickering.

Once I left the garden and went into the house to get a spoon. I heard a frightful noise, my mother crying, talking to herself through gulping angry tears, the kind I knew hurt in the chest. I stood appalled by the living-room door and saw her on the chair before the escritoire with a bundle of papers on her lap. Her back was to me and her shoulders jerked up and down and she slammed her fist on her knee with every sentence. Showers of paper went on to the floor.

"That horrible horrible woman. Everything she did was for herself, nothing for anyone else. I hate her, I hate her. She killed him, the poor old bastard, how he lasted as long as he did I don't know. Selfish snobbish arrogant bitch. Oh God, maybe if I hadn't left he would have lasted longer. But I had to, Father, I had to, I would have died if I hadn't. Oh God don't ever let me be like that, don't ever ever."

I turned and ran softly up the carpeted stairs to the second floor and stood on the landing, and looked through the open door to my grandparents' bedroom, and there on the bed was a mummy wrapped in whitest gravecloths. I screamed and screamed and my mother came leaping up the stairs crying, "Megan, Megan, what is it?" and I could hear William running into the house, and my mother caught me in her arms and saw what my pointing finger aimed at and said, "Oh Megan darling, it's just laundry piled on the bed, I did the laundry this morning. Oh my God, you poor child." When I had calmed down she tried to get me to go into the room with her and touch the sheets and underpants and towels on the bed, but I would not, and I suppose she saw it was no use that way. So she took us downstairs and made us cheese soufflé for dinner, and we had strawberries afterwards. Then when William had gone to bed we sat on the front porch and she tried to explain.

"Megan, I suppose I should have told you before we came, but it's hard for me. I never liked my mother. No, that's not true. I hated her. I know I tell you and William not to use that word, it's a dreadful word, but the fact is it's true. I did hate her. That's why I left and never went back. Not even for my father's funeral. In a way I'm sorry about that, but he was dead, it wouldn't have meant anything to him if I had. But it would have been a sign I guess."

"But why did you hate her, Mum? What did she do to you?"

My mother looked away and then lit a cigarette. "She made me believe that I was wicked."

"Is that all?"

"All? All?" Her voice rose a little and then she subsided and gave me a very gentle smile. I could see wet at the corners of her eyes. "Yes Megan, that's all. It's a terrible thing to do to a child." And then she got me to play Scrabble before I went to bed.

Next day in the garden William said, "Why did you scream like that?"

"Oh, I just made a silly mistake. I thought there was someone on the bed and there wasn't." He did not ask any further.

I think it was soon after that when the jam thing happened. My mother decided to clear out the cellar before moving to the second floor. Down there was an old wringer washer and a quite new dryer, and behind the laundry room was a fruit cellar. William and I had never seen anything like it. A small dark room, almost a cave, the walls lined with shelves. The jars shone in the acid light from the bulb in the ceiling: strawberry jam and plum jam and marmalade, and greengage jam of a beautiful melting chartreuse, mustard pickles and dill pickles and cucumber pickles, mincemeat and apple butter and tomato butter and corn relish, cherries and pears and brandied peaches. Each bore a label with my grandmother's spiky black handwriting, and a date. The earliest was 1947, and the most recent 1965.

"I guess she must have stopped after Father died," my mother said in a puzzled tone. She brought some of the jars up to the clear light of the kitchen and we watched her open some mustard pickle, which William loved.

"The seal's not right. It's gone bad." There was an awful sour whiff and grey-green fuzz oozed up over the circle of white wax.

"What are we going to do with it then?" asked William.

"Throw it out. All of it. I'm damned if I'm going to spend time checking all this stuff to see which is OK and which isn't. Who could we give it to anyway?"

So we did. William and I took all the jars off the shelves and packed them into cardboard cartons. Then my mother laid newspaper over the jars and closed the cartons and tied them tightly with string, and we carried them out to the garbage cans at the side of the house. She looked at the cans and her mouth went tight in a curious pleased way.

"Would that stuff have been good?" I asked.

"Good? Oh yes indeed, she was a very good cook, your grandmother. Yes, very good. A lot of people used to beg her for recipes but she'd never give them out. Typical." She smacked down the last can lid and went back into the house.

At the back of one of the shelves in the fruit cellar I found a notebook bound in black leatherette, filled with recipes in my grandmother's hand.

"Shall we take this home?"

My mother looked at it.

"So that's where she kept it. She would never even show me. Oh for heaven's sake throw it out, Megan." I made for the wastebasket and then she said quickly, "No don't, give it to me." I stood by her as she turned the pages. My Best Apricot Jam. Checkerboard Corn Relish. Chinese Ginger Marmalade. At the bottom of the page was written, "Gordon likes this very much."

"Who was Gordon?"

"Your grandfather," said my mother absently.

"Are you going to make some of those things for Daddy and us?"

"What?" she snorted. "Maybe I will at that. She'd hate to think of him eating anything she'd had a hand in."

"Why?"

"Oh she never liked him, Megan. Never wanted me to marry him."

"But *why*?" I did not understand why anyone would not like my gentle cheerful father, and I was getting annoyed because my mother was not really listening to me. She went on turning pages as she talked.

"Maybe I should say instead that she disapproved. Because he'd been married before. A marriage to a divorcé wasn't really a marriage. Something like that. No, that's not altogether it either. Strawberry Rhubarb Jam, I remember that, it was damned good. No. The real thing was, she called it 'getting another woman's leavings.' I suppose she despised me. As usual."

"What are leavings, Mummy? Mummy?" She came out of the notebook then, startled.

"Oh Megan, I've said all sorts of things to you that I shouldn't. I'm sorry dear, now take the book and put it in that Vancouver box in the living-room."

The weather was now extremely hot, far hotter than it ever gets in Vancouver, and William and I found the cool green shade behind the maple tree more of a refuge than ever. Because of the heat we developed some new dramas, particularly one about being castaways on a desert island in the South Seas and finding treasure there. Then somehow this got merged with the play about the old people, for William got the idea that we should bury something under the maple, leave something when we died.

"But who for? We're going to be the last, there won't be anyone to find it."

William was stubborn, and said perhaps millions of years later new kinds of creatures, people, might grow up upon the earth, and they would find it and spend centuries trying to figure out what it was and finally would.

"Archaeologists you mean."

"Yes, that."

I did not think this made much sense, but I liked the idea of making our own buried treasure, and so we set about finding a suitable place and thinking of something to put in it. We took a couple of the big dinner-service spoons without telling my mother, and dug about among the roots of the maple. I heard William say, "Aaah!" in fright. He was staring at the earth before him and had gone pale so the freckles on his nose looked dark.

"There's something here, Megan. Look. Someone's buried something here before us."

There was a metallic strip shining in the black dirt. We waited a moment—I know I imagined that a bony hand might be holding whatever it was—and then we began to scrape and shove with our big spoons. There it was, a cigar box, with a picture on it of a woman in a long flowing red dress. It was wrapped about with yellow Scotch tape, and dirt lay deep in the long cracks between the strips of tape.

"Let's open it, here, the tape starts here!"

"No no no!" cried William, and he got up and backed away from me, holding the box to his chest. "No no. It was my grandfather's. We can't open it yet."

"What do you mean, not yet? Come on William, I want to see what's in it!" But he began to cry, and I did not want our mother to come out, so I said, "Oh all right, well what do you want to do with it then?"

"I want to leave it till the last day, the day before we go home to Vancouver."

"But why? Whatever for?" But I could see it was no use. So we put the box back where it had lain for who knew how long, and covered it with earth again. Then every time we started to play behind the maple tree we made a little ceremony of checking to see that the box was still there.

"What did Grandfather die of?" I asked my mother.

"Something called arteriosclerosis. It's a disease that makes your arteries harden, and after a while you can't walk any more. And with a lot of people it makes their brain go funny too."

"Did that happen to him?"

"I think it did, at the very end. He was ill for years. Just gradually getting weaker and weaker. The last couple of years of his life I don't think he even went downstairs any more."

I imagined an old man sitting in William's room, looking out the window at the roses and the maple tree; sitting and looking, sitting and looking; watching her do what he told her to, in the garden; thinking about the box buried under the maple?

"What did he look like?"

"Goodness, why are you so interested in your grandfather all of a sudden? Here's a picture of him. With your grandmother. I'm taking this one home with me. God knows why, but I am. Here William, you come and look too."

My grandmother sat in a garden chair in front of the maple tree, looking up at my grandfather who stood beside her. Hers was a strong face, aquiline nose and piles of dark hair. She looked as if she had just spoken and was waiting for an answer. He was looking past the camera, perhaps trying to evade the sun, which glinted off his glasses so his expression was not visible. He was smiling faintly.

"He looks nice," said William with satisfaction. "Why is she holding him like that?" My mother and I looked and saw that her left hand was clasped around his wrist.

My mother humphed. "Probably just finished telling him to do something. Or just starting to tell him to do something." She turned the picture over. There was that familiar handwriting: "Gordon and Louisa."

"Why did she tell him to do things so much?" asked William.

"Oh William, you do ask the most impossible questions." My mother sighed. "Well, try to understand. She was what's called a managing kind of person. Do you know what I mean? She was always wanting to run things and get them organized and be in control. My father used to say she should have worked in the bank instead of him, she'd have ended up president."

"Did he?"

"Oh God no, he never made it past manager of a small branch I don't think."

"Well couldn't she have worked in a bank too?"

"Oh William." My mother was getting impatient. "She wasn't like that, things weren't like that. Yes, sure she could have worked in a bank, or somewhere else. Your grandfather wouldn't have objected. In fact I think he would have been glad, and it would maybe have taken some of the heat off him. God knows it would have off me. But she didn't believe in women working outside the home. She thought women who did were low-class, do you know what I mean? Even if they had jobs like being a doctor or a lawyer or something. She thought they were failures. Failures. Even if they were married too. Does that make sense to you, William? To you, Megan?"

I started to speak but my mother had got all wound up and went on.

"So she could have, but she didn't, she stayed here in this house, and controlled every inch of it. There wasn't a single thing my father or I could do or say that she didn't know about and criticize and oversee. She was like a, like a searchlight in a little room. She ate us up, at least she ate up my father and she tried to eat me."

"*Eat* you?"

"Oh William darling, not like that." She took William on to her lap and held him close.

"You don't have a job either, Mum," I said. I had not really thought about that before. My mother looked sharply at me.

"No I don't Megan. It's different, though. I mean the reasons are different. I like being at home with you. And this fall when William's in school all day I'll start back. I'm going to take that refresher course for nurses, you know that. And likely by next spring I'll have a job in some hospital in Vancouver. But I wanted to be home with you both when you were little. I've loved it." And she reached for me and held me with her arm while I leaned against her and looked down at William's nice brown hair. I could feel her trembling. "I hope you've been happy. I hope I've done right by you. Being here makes me feel I don't know what the hell I've done or why I've done it." William kissed her and then she really did cry for a bit, and said, "I've wanted it to be different for you, so much I've wanted that."

The *National Geographics* had all gone to the Salvation Army, and so had the closetful of my grandfather's heavy dark clothes. My mother was giving his study a final cleaning. There were stains on the wall by the window; I had worked it out that his chair had stood there, and his hand had rested on the wall while he looked out.

"Who took care of my grandfather while he was sick?" I asked.

"Why she did, of course," and my mother sounded surprised. "Your grandmother. Who else?"

"All by herself?"

"Why yes. Until the end, they had a nurse then, for the last few weeks before he went into hospital to die. I suppose she couldn't cope any longer. She wouldn't have anyone before that. Didn't want anyone in her house." Her voice was very cold. She made a few last scrubbing motions at the stain. "Well, I can't get that out. Whoever buys this place will paint it anyway. All right. That's done." Only the mattress on the floor remained. William's clothes were stacked in cartons, or in his open suitcase along with the cigar box. My room looked much the same, except that there was a pile of L. M. Montgomery books next to my mattress. My mother had given way and said I could take these as well as the little ballerina. William did not even make a fuss over this. He was more and more preoccupied with going soon to the Museum, and with opening Grandfather's cigar box. We played our play by the maple tree still, but William kept breaking out of it to talk about these approaching events, and that spoiled the continuity.

The guest-room my mother used was also almost bare, and that left only the master bedroom with any furniture in it. The mahogany double bed and the matching dresser with the swinging mirror were marked for auction. The dresser was empty. I had watched my mother take the stuff out—long peach satin nightgowns with what looked like bags for my grandmother's breasts; things I thought were bathing suits but my mother said were foundation garments, white with lace around the tops and smelling of Javex; stacks of packages of seamed nylons. Out of the clothes closet came dresses, somehow heavy in the same way as my grandfather's suits, though of course the fabrics were different. Greys, browns, blues, dark greens. My mother held one up to her, and it came almost to her feet.

"Was my grandmother bigger than you then?" William asked.

My mother laughed shortly. "No. It's just that the styles she wore were longer. Even when skirts went up she never changed. Said they weren't ladylike, like that."

"Why do you talk always about her and not about him?" said William crossly. "You never say anything about what he was like."

"Don't I?" My mother set down the dress she was folding and looked at William. She frowned, not angrily. "Well, why don't I? I guess because he was gentle, not like her. Yes, he was gentle, and he didn't say much. But I loved him a lot William, you must know that."

"Did she love him?" I asked, picking up a blue flowered nightgown.

"Put that down dear, for heaven's sake. I've just finished folding it. Did she love him? Oh dear heaven." She sat very still. "Yes," she said slowly, regretfully, "I suppose I would have to say that she did. In a way.

But Christ, what a way." She shook herself. "Now Megan, get that box off the shelf in her closet and put it in my room."

"What's in it?" I got it down; it was the last thing in the closet.

"Letters."

"What letters?" asked William.

"Just letters, that's all. Put it on the floor by my bed." I did. My mother's tone made me uneasy, for it was not one I knew, and although I wanted very much to look at those letters I did not dare to. Somehow I did not think my mother was looking at them either. In fact she used the box as a night table, putting her ashtray and whatever book she was reading on it before she went to sleep.

Time was now getting jammed together, and we found all sorts of places we wanted to go to and things we wanted to see in this last week of our stay. My father said that the kittens were now trying to get out of their box, and within a few days we would be home and could see them. I could hardly wait, but William was not very interested.

We went finally to the Royal Ontario Museum, and it was a terrible disappointment to me. Oh, I was fascinated by lots of the things—the Amphorae (I had not realized they were so big), the dinosaurs, the mummies, the stuffed animals, the great glass cages in which were life-size models of native Indians caught dead-still weaving baskets and pestling corn, the echoes of the hall lined with Greek statuary, whiter than I had ever imagined. But, except for a few large free-standing objects, I could not *touch* the things. I had thought the cases would open, that we could take out the combs and mirrors and jewellery and hold them, turn them over, and feel what other hands had felt, hundreds and thousands of years ago. Where do children get such ideas? The frustration was terrible. I can think of few other times in my life when I have felt so blocked, prevented. I can see my hands yet, flat against the glass guarding a dozen little Chinese horses, and how I longed to feel the hard delicate hoof in my palm.

William did not mind as I did, and walked about with intense enjoyment in everything. Then he found the thing he liked best, and wanted to go back and back to it till finally my mother said he could just stay there then, and we would go and see the dinosaurs and the Roman coins again and come back to fetch him when we were ready to go home. His exhibit was in a section where there were a lot of stuffed animals; I remember polar bear and marten and a great eagle. A glass case hung on the wall, and in it was what looked like a cross-section of two steep hills with a valley between. Down the hill on the left were arranged hundreds of tiny skeletons, some very little animal like a shrew? a deermouse? all as if running for their lives to get down into the valley and across to the other hill. But their numbers decreased as they went down the steep slope, and in the depths of the tiny valley there were only a few skeletons running along in a line, and then finally on the rise to the other side there was only one solitary set of minute bones, so small you could hardly imagine there had once been a heart beating under the thimble-size rib cage. My mother read the accompanying legend to William about five times before we left him there, something about Nature's way of ensuring the survival of the species in the face of illness and disaster and attack. When we went back to get him he was standing quite still, gazing, and the freckles on his nose looked very dark. He was quiet that evening, but when I reminded him that next day we would open Grandfather's box he smiled at me.

We waited until after lunch. My mother went upstairs to finish packing, and we went out to the dark green space behind the maple tree and began to dig. My mother had found the two big dinner spoons missing when she came to prepare things for the auctioneers' truck, and had scolded us sharply and given us two old wooden spoons from the kitchen instead. They were actually very good for digging. We got the cigar box out, and I let William open it. No, really I was afraid even to suggest that I might do it, being older, because his face was so intent. So he got the end of the tape free and unwound it, breathing noisily, for he was getting a cold. As he unwound I saw that Grandfather had not made a perfect seal around the box, and I suppose that should have prepared me. When he had got all the tape off William sat there, he sat there holding the box on his lap.

"Oh go *on*, William, open it."

He did, and there was a horrible mass of squirming bugs and worms and soggy dirt and wilted paper, and William cried out and threw the whole thing away from him on to the bare earth, and everything fell out in a heap. We both got up and jumped back, and William began to cry. I was afraid my mother would hear, so I worked as hard as I could to calm him and finally did, though he was very pale.

"If we touch the papers carefully with the spoons we can scrape off the yuck," I said, "and then maybe we can look at them."

We squatted a couple of feet away from the mess and stretched out our spoons and began to scrape. Soon the pile of insects and worms and dirt began to look just like that, insects and worms and dirt, and we moved closer to the bundle of papers and spread them out on the ground. But it was no use, the damp had done its work. The writing—pale blue, sloping—was all sodden and blurred, and although the papers were clearly letters we could make out nothing. We were working with our hands now, turning over page after soggy page, trying to find something that could be read.

"Can you read that? Can you read that?" William kept asking. But I could not. And then I found one small note which I suppose had been tucked into the middle of the pile and so escaped some of the wetness, and on that I could read two words and two words only.

"Darling Rachel," I said.

"Darling Rachel," William repeated, giving the words no meaning. He looked up at the house and screamed. I looked and saw at Grandfather's window a human shape, and then my mother opened the window and called, "Children, what are you doing? Megan, what's wrong with William?" He had fainted, I knew that was what it was, he had done it once when he was getting flu.

"But who was she?" said my mother, weeping, as she stood before the kitchen sink, burning my grandfather's letters to Rachel. We had got William into bed, and she had laid out the letters over the kitchen counter to get them dry enough to take the flame, and now was picking them up one by one and burning them. Periodically she ran the cold water, and the grey-black shreds went down the drain. "Who was she? And why did she send the letters back to him?"

I did not answer, wishing only that she would stop crying. I tried to make myself think there was something romantic in it, love-letters buried for years under a tree. I had just finished *Kilmeny of the Orchard* and thought perhaps that L. M. Montgomery would have approved of such a notion. But the nasty mess in the box and William fainting and my mother's tears did not fit, would not go into the shape. I did not know what to do. Finally she made some lemonade and we sat together on the steps of the front porch and drank it, looking out at the square tidy park. The water fountain wasn't going any longer, the city was trying to conserve water or something. My mother sat very close to me with her arm around my waist. All that evening she was very gentle to us. William got up for supper, his nose streaming with cold, and afterwards we watched a program together about game reserves in Africa. I looked over at my mother in the semi-dark and saw that she was crying silently, and did not look at her again.

The next day the lawyer came to the house and my mother gave him the keys and we left. We stayed that night in the Royal York Hotel, which was wonderful for William and me. When I woke next morning, my mother was sitting by the window, looking out at the lake. She was drinking coffee and smoking; she never smoked before noon. On the table before her was the box of letters from my grandmother's closet.

"Please now won't you tell me?"

She took me on her lap, which she rarely did now that I was so big.

"I told you before Megan. It's letters." She sighed. "Letters your grandmother and grandfather wrote to each other when they were"—she paused—"when they were young. And all my letters to them, and all my letters to her after he died."

"What are you going to do with them?"

She sighed again. "I'm taking them home, Megan. At least, I'm going to mail them, the box is too big to go in my suitcase and we have enough stuff to carry on the plane as it is." She turned me on her lap so our eyes met, and it was as if her glance was coming from very far away. "I've been thinking about these all summer, dear, trying to know what I should do. I thought finally I would throw them out, but now I've decided I have to read them. I've not been able to do it yet, but I will. Your father will help me, I know. Oh my God it's going to be hard, but I have to. I have to try and understand what was going on in this house all those years."

"But won't that be interesting, to read all the letters you wrote way back then? Like reading a diary?"

"At least in diaries people tell the truth. Or I guess they do, I've never kept one. But what I wrote to them, what I wrote to her—well, it was mostly lies."

"You told them lies?"

"Oh not like that Megan," and my mother set me down and took another cigarette and walked back and forth in front of the view of Lake Ontario, not looking at me or at anything. "Not real lies. I just didn't tell them what was really going on in my life. What I was really like. They were sort of form letters. Dear Mum and Dad. Dear Mum. Years and years of form letters. Dear Mum. Why she kept them I do not know. I truly do not know. And as for those letters they wrote to each other . . ." I thought she was going to cry again, but instead she went over to William's bed and shook him and said, "Hey come on William, do you want us to miss our plane?" And then she phoned the desk to see if the hotel could provide us with wrapping paper and string, which it did, and then we had an enormous room service breakfast and counted all the boats we could see on the lake, and five planes, and then we left.

On the way to the airport my mother got the taxi to stop at the main post office and we all went in. She registered the parcel, and then William begged to be allowed to deposit it in the mailbox with the special large opening for big packages. We watched him walk away across the great shining marble floor, and my mother smiled, I suppose because William looked so serious and responsible.

William was sick on the flight home, but by the time we reached Vancouver he was feeling all right again. It was wonderful to see my father. The kittens were tumbling on the rug and trying to catch dust in the sunshine. William and I played again in Burrardview Park with the blue wind rushing in off the water and the mountains brilliant to the north, and I was very happy.

But the parcel of letters did not come and did not come and did not come. My mother waited and waited, and then got on to the post office. There were tracers and searches and angry telephone calls and letters. Finally, towards the end of November I think it was, she recognized the hard fact that the parcel was not going to come. She cried harder and longer than I had ever imagined a person could, and my father was half sick with worry over her. It was terrible, and William got bronchitis again.

Then my mother pulled herself together. She wrote her Christmas exams for her nursing refresher course and passed them well, and in the New Year she got a job at the Vancouver General and things were better again. And it was not until many years later that William finally told me what he had done; that on the walk across the gleaming floor of the Toronto post office he had licked his fingers and then smeared and smeared across the address and the registration slip until they were pale and runny blue and quite illegible.

Beatrice

My parents died stupidly in a car crash when I was fourteen. No one was surprised to learn that the fault was my father's. I already knew he was ineffectual, that he and my mother were soft and blurry instead of clear-edged and definite like the parents of the popular boys at school. I would watch as these beings emerged from their glossy cars on prize days and at half-terms and strode to meet their offspring, and then would see our old Morris coming slowly round the curve of the drive, my father peering through the windshield—of course the wipers were not working—and looking as though he had never visited this part of England before. Their deaths and my grief were a drama in the school. "Speeding, I suppose?" the boys asked admiringly, when such questions could decently be put. I said, "Yes," because I could not bear the truth, that my father had caused the crash by driving too slowly. Soon however the drama heightened. I was to leave the school, leave England, and go to Canada to live with my mother's sister, my Aunt Beatrice. The éclat of this news carried me through almost to the end of term when everyone lost interest in Hugh Proctor for the simple reason, apparent to all except myself, that I wasn't going to be back and they were. I felt even worse than just after the funeral, yet I never dreamt of trying to get out of going to Canada. Arrangements had been made. I was most miserable. In this state I boarded the plane that flew me to Malton Airport near Toronto in August 1959.

Nineteen years ago. More than half my life. Perhaps more than half my life is over now? Do workaholic lawyers make old bones? Do not-quite-alcoholic lawyers? My daughter, my daughter, think of her, at the very beginning of her life. Think of her lying up there on the fourth floor of the hospital, dark-haired, rosy in her rosy blanket in that odd plastic container they put new babies in. Imagine being able to say My daughter. (Do you have any children? Yes, I have a daughter.) I wonder if my parents felt anything like this when I was born. If they did I wasn't worth it. I have a feeling my mother would have liked a girl. She always loved her sister Beatrice very much. Admired her. "Beatie was always the one for doing things. We all thought it was so brave of her to go off on her own like that, after the War. Get this war over, and I'll be 'off to Canada to have adventures,' she always said. And then she really did it." How many times at home did I hear variations of this speech.

Aunt Beatie. The first clear understanding of her, after a month or so of dislocation and muddle in the new country. I am in a large meeting hall. Hundreds of people on cheap wooden chairs. Banners. Posters. Too much smoke, too few ashtrays. In front of the platform stretches a huge hand-painted paper banner reading Fair Play For Cuba. On the platform a fat sweating young man is describing the evils of prostitution in pre-revolution Havana. This is not the first political meeting I have been to with Aunt Beatie. Already in one month with her I have heard more about politics than in the entire rest of my life, but still the main point has simply not sunk in. Tonight it does. The young man says quietly, "Now all that is changing," and these plain words make the crowd clap and cheer. But those people in Cuba are Communists, I think. Then these people are too. Then Aunt Beatie is. But she can't be a Red, she's my aunt. I turn and look at her. My body trembles with the power of what I have finally understood. She is folding her hands in her lap again, smiling gently. She could be a teacher at a girls' school. Curly pepper-and-salt hair, neat features, pale skin, glasses; a dark blue cotton dress, white summer sandals; composed, reserved, calm, what the hell is the word to describe that I-am-my-own-person air? There she is, and everything about her says she should be at the village concert to raise money for the restoration of the Norman arches in the church. But instead she's here cheering for some pinko peasants in a hot messy little country somewhere off the coast of the States.

Not that I expressed all that to myself then. All I could get clear that night was that I was in the charge of a Red and did not know what to do. I remember thinking I should run out of the hall and go to the police. All of which provides illustration of the subtle and complex ways in which ideology is transmitted. My parents had never discussed politics. I do not remember anything that could remotely be described as a political discussion at school, nor a history lesson which either (a) went beyond the Wars of the Roses or (b) dealt with those wars in anything except a first-there-was-this-battle-and-then-there-was-that-battle way. Yet from some unseen nipple I had sucked and absorbed and made my own every single conservative notion a fourteen-year-old boy could possibly have. I was profoundly shocked, the more so because Aunt Beatie was a woman.

Of course I did not go to the police. I stayed where I was, and in fact I lived with Aunt Beatie for over ten years.

I would be embarrassed to say how long a time went by before I gave any thought to the effect my sudden eruption into her life must have had on Aunt Beatie. My other relatives in England and the family lawyer had told me that I should be grateful to her for offering to provide a home for me. I obediently thanked her shortly after my arrival. But it still seemed to me that she had simply done the correct thing and there was nothing to be so impressed by. After all, I was only fourteen and could scarcely live on my own, could I? And she was family, wasn't she? After a while around the revolutionary party and around Harbord Collegiate, I learned that these to me logical sequences of thought were not at all so to others. Fourteen-year-olds could and did live on their own and blood ties meant very little to some people. These however were what I thought of as people with funny names. Thinking of them that way helped me to counter my sense of utter alienation. An uptight middle-class English schoolboy, late fifties model, dumped into the polyglot polychrome ferment of a downtown Toronto high school whose high standards were the springboard out of the ghetto, on to the campus, up the hill—I would have found Mars easier. I see I have again wandered away from Aunt Beatie, unable to resist talking about myself. Yet it was she who enabled me to make it at Harbord. I think she thought it was her responsibility. As a matter of fact I think her ideas about her responsibility for me weren't that different from mine, for all her revolutionary politics. I didn't know then how hard it can be just to do the correct thing.

Serious revolutionaries work very hard. Most have jobs or families or school or some combination thereof, and on top of that have a load of political activities that would break the back of a Percheron. Aunt Beatie got home from work (she was the supervisor of a pool of forty typists at the telephone company) around five-fifteen. We had dinner at half past. She would then leave for, say, a six o'clock executive meeting of the revolutionary group she belonged to. (Our, her, apartment was west of Spadina and north of College, near her group's headquarters—also near those of many other left wing groups, for they all seemed to hole up in the same part of the city.) This meeting would last until perhaps half past seven. She and others would then leave for the political assignment/event of the evening: union meeting, branch meeting, outside group meeting, film series, lecture series, picket, fundraiser, whatever. She would get home about eleven and work for an hour or so before bed—revise an article or speech, do the books for some organization, read a couple of Marxist papers. We got up no later than six-thirty in the mornings, and sometimes Aunt Beatie would rise earlier to finish some task. This went on five days a week. Weekends were sometimes lighter, sometimes a lot heavier. She had to attend two-day conferences or series of classes, or sell the revolutionary press, or go on sub-drives. So I suddenly had to be included in this jam-packed schedule.

At the beginning especially I must have been an awful chore. I could dress myself and make my own bed, period. I had never cooked, cleaned, shopped, laundered, budgeted, and I despised these kinds of work because women did them. I truly thought Aunt Beatie was joking when she told me to set the table, turn on the oven, take out the garbage. Disbelief phased into outrage and there was confrontation. Of course I hadn't a hope against Aunt Beatie. Any woman who had been organizer of a large urban branch of a Trotskyist organization throughout the Cold War would have made short work of me and my adolescent sexism, and Aunt Beatie did so. By Christmas of 1959 I could prepare an adequate dinner and iron my shirts quite well. By the time I left high school I was an accomplished houseworker and in fact I made extra money in university by doing housework. I don't do it now, of course, Robyn doesn't like me to. It makes her feel funny to see me cleaning the floor. I know why, but we don't talk about it.

So Aunt Beatie had to train me as a co-inhabitant of her apartment. That took up her time and energy. She had to re-educate me about education, and so did that. I thought school was silly and boring. In my defence I will say that I can't think of a more awful bunch of teachers than those at the school in England my parents had mortgaged themselves to send me to. She also had to try to teach me to take my life seriously. I had no thought of what I might do when I was finished with school. I had no special bent to direct me. I suppose I had thought the grownups would suggest something when the time came. Aunt Beatie would not tolerate this aimless drifting. She made me talk about what I was learning in school. She argued with me. She suggested things I should ask my teachers. She criticized my papers after they were handed back to me. (Not before, because if I incorporated any suggestion of hers it would mean that the paper wasn't wholly my work and therefore the mark I got wouldn't really be mine either.) She got it into my head that as a minimal goal I must learn to do some sort of saleable labour so that I would not be dependent on anyone. She was of course firmly in favour of all forms of social assistance, unemployment insurance, etc., but to work was clearly best. I don't know whether it was the Marxism or the North Country-ish self-reliance which said that more strongly in her. It doesn't matter I suppose. Though it would have, to her. She hated political sloppiness.

Have I made her sound cold and strict? There was more than that. Once, soon after I arrived, she made us pancakes for Sunday breakfast. I had never seen or eaten such things and said I would rather have toast. I'm sure I said it in my snottiest oh-you-colonial tones, too, but she got the underneath sense. My homesickness and grief made this unfamiliar food just too much for me to take that morning. She put the batter in the fridge, kissed me warmly, and made me a pile of toast. Then she got me to take the toaster apart and clean out the crumbs, and I felt much better. Later on in that hard first year in Canada, at Harbord, I was struggling with Latin. In England I had learned to decline nouns using a certain order of cases, and in Canada another order was used; it sounds a small thing, but for a kid at school it's not. I kept making stupid mistakes in my translations because of this change. Aunt Beatie came home unexpectedly early one evening from a meeting and found me crying over my homework. She ignored my humiliation and got me to explain the problem. We discussed it while my face dried. The act of explaining helped. Time of course did the rest. A few weeks later Aunt Beatie said, "You know, Hugh, there's no shame in a man's crying. The tear ducts are presumably there to be used, eh? Crying's sometimes a good way to clear out your feelings. For man or woman." And she went on to something else.

I cried this morning when my daughter was born. I am wordless to say what I felt when she emerged from Robyn in that final quick bloody rush. She lay, small yet substantial, in the doctor's arms, gently moving her hands and feet. She gave that low cry and her eyes looked right through my head and past it through to some realm I don't even know about. What can you do with emotions of such power but let them have their way, and cry as they possess you?

That Aunt Beatie had an emotional and sexual life never occurred to me either. Like most teenagers entranced and obsessed and agonized by their emerging sexuality, I was revolted by the notion of anyone over perhaps twenty-five engaged in sex. Aunt Beatie was in her mid-thirties. She was my aunt. She was not married. Three taboos. If I'd been older I might have wondered how on earth she would find the time to get involved with anyone, or speculated about the quantities of men available who would appeal to a woman revolutionary. But I wasn't and didn't; so I was shocked one night when I got up to go to the bathroom and glanced in at Aunt Beatie's open bedroom door as I went past. There she was in her bed all right, but not reading as I had expected. She lay naked, with her back to me, in the arms of an equally naked comrade who had been at the meeting held in our living-room earlier that evening. They did not see me. On the way back from the bathroom (I did not flush, because of the noise) I stood by her door, terrified lest they should notice me passing it. I considered lying on the floor and crawling past. My embarrassment then achieved indescribable heights as I realized, from the passionate (though muted) sounds they were now making, that I could probably do the Highland fling in the doorway and they wouldn't notice. I slipped past, and back to my own empty couch. Aunt Beatie's was the first live unclothed female body I had ever seen. The curved pale line of hip-waist-back affected me powerfully, and I lay awake and active for some time.

Next morning at breakfast Aunt Beatie was calm and cheerful as usual. Comrade Stan ate his porridge and boiled egg placidly at our table as he had frequently done before. I had assumed that he and indeed any other visitors we had did their sleeping on the couch in the living-room; I felt much older than the day before and at the same time I felt silly and inexperienced. And I felt wicked because I had been aroused by my aunt. I got off to school as fast as possible. I think I was then sixteen.

Perhaps it was this incident that led me to look consciously at the people with whom Aunt Beatie spent her life when she was not at her job. To this point they had been simply "the comrades," a group of adults who were part of the backdrop of my life. The foreground increasingly was filled with other kids at school and by school activities I was finally getting into, having lost the worst of my English accent and having gotten rid of my ties and grey flannel trousers. I was an all-right student, not terrific but more than holding my own, and I had a mild aptitude for debate which a fanatical coach was insisting that I develop. I looked forward to school now instead of dreading it. The idea of going to university was becoming interesting. But these comrades now, these people with whom my Aunt Beatie evidently had far deeper and more complex relations than I had understood, what were they really like? I passed them in review. I made no discoveries about their respective sex lives like that I'd made about Stan's, but I did reach one very solid conclusion: my Aunt Beatie was the best of the bunch by a wide margin.

The women—remember this was the very early sixties—generally fell into two categories: the housewife-helper of the bigshot Marxist, and the socialist sexpot. The men were more varied (there were far more of them of course). Thin hungry-looking office workers. Pudgy brooding students. Elderly men with dried skins who walked softly so as not to wake the black-lung cough. Downwardly-mobile envoys from Rosedale: these affected strange raiment, black capes and such. Big construction workers, their hands permanently crusted with chalky stuff. Literary types who always wanted to have forums on Christopher Caudwell instead of on the latest CCL convention.

Aunt Beatie's singularity and decorum set her off from the other women, and she was straightforward and truthful in a way that most of the leadership males weren't. There was no bombast in her. When she said, "I hate the bosses," she meant it literally. She loathed her own boss—he was fifteen years younger than she and earned a third more—and she loathed all bosses. The passion with which she said these things was striking in the context of her general reserve.

When the women's liberation movement erupted into the branch, new kinds of women began to join. I suppose that for the first time in her years of commitment to the revolution Aunt Beatie had comrades she could really talk to. She was intensely admired by some of these younger women. They made her almost a matriarchal cult-figure for a while. They all wore jeans or pants, of course, and one day Aunt Beatie appeared at the hall wearing a neat denim pantsuit. They literally cheered her. "I've wanted to do it for years," she said, smiling. (Later she led and won a battle at work over wearing pants on the job.) After that she wore skirts and dresses only. Of course the political crush on her did not last. (Everything contains within it the seeds of its own destruction.) She was found to be sexually prudish, secretive, and too loyal to male organizational structures. All that came later, though.

One of the few issues on which many comrades openly criticized Aunt Beatie was her treatment of me. (Of course, no one spoke ill of her directly to me; but I overheard, guessed, saw enough to know that my aunt was ruthless in internal party struggle and that numerous members of the branch bore the scars to prove it.) They thought, and said, that she should be actively encouraging me to participate in revolutionary politics—join the youth group, write for the paper, and so on. There was a rough faction-fight in the branch during the mid-sixties (again and once again the "question of the NDP"), and I believe that some comrades then rumoured it that Aunt Beatie's own commitment to the party must be weakening, for otherwise would she not urge me to join? Some of the comrades were themselves red diaper babies. As infants they had slept on registration tables in meeting halls while their parents spoke and argued and raised funds for this and that leftist cause in the late thirties and during the war. As children they had helped to run off innumerable leaflets on the Gestetner in the basement. Now as teenagers or young adults they were getting their own files opened by the RCMP and finding their own way around the changing left in Toronto. They couldn't understand her refusal to pressure me. Aunt Beatie's reasons were as usual clear in her own mind.

"Hugh, your job now is to get through school and get trained at something. I've no patience with these young ones who're up half the night selling the press and then fail their Christmas exams. And then what? All to do again, or else sponge off their parents for years. Young workers, that's different. I wish we had more of them, the branch is middle-aged, mostly. But you do your work. When you're through there's time enough to get involved in the movement, and you'll be some use then because you'll know how to apply yourself. If you decide for the revolution, that is." But of course she never doubted that I would. And although I didn't join the youth group I was a regular at the party hall. I helped stuff envelopes and do phonings after my homework was done. I read two or three Marxist papers a week. I went round with the collection buckets or sold buttons at rallies on the weekends. Being a radical at Harbord was not remarkable. There was even a certain cachet to it, useful with girls.

I wonder what I will do about my daughter. Will she hate me when she grows up for having pushed her in one direction or another? Is it possible not to push? Is it desirable? Even if you don't actively push, isn't it obvious that given your inescapable example the child will get a very clear message about where you think the right road lies? But I don't want her on this road. I can't bear it if she hates me. I didn't know love could erupt this fast. Yesterday she was simply The Baby, and now she herself is here and thinking of her makes me ache. Aunt Beatie never had a baby. I certainly never hated her, although I have felt very miserable because of her.

She approved when I decided to go to university. Money for this was not a major problem because of the insurance arrangements my father had made. Typically, they were muddled and had taken some sorting out by the lawyer, but they paid for my education. Off I went to U of T in the fall of 1964. My four years there saw the rising of the student and black civil rights movements in the States, the radicalization spilling into Canada, the beginnings of the movements for women's liberation and against the Vietnam war. Big things began to happen inside Aunt Beatie's group and in all radical organizations in the country. She was busier than ever, and so was I. I took an honours degree in history, and finally "did" the Wars of the Roses in a way that made some sense. I was on the UC debating team and then on the Varsity team. I began to think about being a lawyer.

It is hard to explain my political life during those years, yet important to me; there are distinctions I must make. I was not inactive. I went as usual to all the major events Aunt Beatie's organization put on: the Russian Revolution banquets and the May Day banquets, the New Year's parties, the Labour Day picnics. I went frequently to the weekly forums. When leaders of the Fourth International from Europe or South America came on speaking tours I went to hear them, and I went to all the major demonstrations on the various issues which those turbulent years cast up. I see that my choice of words has said it all: "I went to." I was not part of, not involved in. I don't think Aunt Beatie saw that. There were more things happening, so I turned up at more things, so she took that as a deepening commmitment.

I see that I have reached that part of my own life story which I find so unsatisfying in so many biographies and autobiographies: the changeover from youth to adult. The years of early and middle childhood are always clearly described, for the colours of those first designs on the mind and heart never fade. Things get more complex in adolescence. Then suddenly there is the grownup, saying and doing things that sound boringly like the things all other grownups say and do. I cannot in my own mind pinpoint a month or even a year in which the shift took place. All I know is that when I left high school I took it for granted that after some further education I would move into radical politics in a committed and active way, and when I graduated from university I knew, submergedly, that I would not do so. I did not yet verbalize this to myself. Later, I found out that the comrades' analysis was that my convictions had been sapped by petty-bourgeois academia. They thought that my history and political science and economics professors, conservatives and liberals to a man (and come to think of it, they were without exception men) had overwhelmed my elementary understanding of Marxism with the sophistication of ruling class ideology. I was defensively angry when I first heard this. Then I realized that it was after all a very commonsense explanation for my change, so I considered it carefully. I still don't think that was the reason.

No. That's too cerebral. On the level of argumentation, theory, analysis, I could, I can, still say that the Marxist world-view makes infinitely more sense than any other. When I go into the courtroom the class lines are so clear to me they're like a coloured overlay on a page of print. Sometimes in the courtroom I'm on one side and sometimes on the other, but those lines are burning clear. But. I think what was missing in me was some element of conviction. Confidence? Maybe that's the word, confidence. Aunt Beatie had so much of that. She didn't rely on other people's approval. She did what she did, she was sure. I wasn't. For all I *knew* about the potential of the working class, about strategy and tactics, the contradictions within capitalism intensifying to the point of rupture, the links between and among the various groupings of the oppressed, the role of the vanguard party, the role of the trade union movement—I still didn't hold it in my heart of hearts that it would *work*.

I can hear everything you would say to that, Aunt Beatie, but I'm sorry, it's me that wants to say my say now as I never did to you. Well then, why didn't I have that confidence, that conviction? Why you and not me? Does the root cause wriggle all the way back to that boy in the English school feeling inadequate because his parents were not like others? Well, back further, why did he/I feel that way? Or does it go back to the night I learned that Aunt Beatie was a Red and I *didn't* run and fetch what I didn't then know was the class enemy, a cop, to save me from her? Action, being active. That was Aunt Beatie but it is not me. I am more of a re-actor. Is that a middle-class trait, finally? Is there then a class-based explanation of why my feelings (not my head) were inadequate for the revolution? Or is all this just a fancy rationalization of political cowardice?

I've spent a lot of time and gone to the bottom of a great many bottles trying to find out, but I still haven't.

In some ways I am rather pleased with what I did do when I finally admitted to myself that I was not going to join the revolutionary party. Unlike many people I can think of, including some so-called radical lawyers (there is no such animal in my opinion), I have not pretended. I have not clad myself in trendy dishevelled garb and set up shop in a downtown reno and boasted about "only" charging two-thirds as much as straight lawyers do. No. I went through law school working as hard as I could, and did well enough so that I first articled with and then joined one of Toronto's more prestigious firms. Not obviously one to which birth in the right family is the door-key, but a very good one. There are quantities of partners, all with solidly WASP names like Galbraith and Twining and Mortimer, and I expect that Proctor will fit in very nicely in a few years' time. I wear three-piece suits and dark socks. My hair is neither so short as to make me appear sexually repressed nor so long as to look fashionable. Until Robyn and I were married I lived on nearly the top floor of an expensive high-rise at Bloor and Yonge. We now have a smallish house in the nether reaches of Rosedale. I know the block I want to live on in Forest Hill, and presumably at some point we will do so.

In fact I suppose that I have become almost exactly what my parents thought they wanted me to be. They cannot have imagined, though, that I would subscribe to half a dozen radical papers from around the world or that I would be known to many lefty groups in town as an almost sure hit for money every fund drive. (I won't let my name be used publicly as a donor or endorser, though, and there are now a few groups to which I will no longer give.) What a joke it is, what a joke I am. I wonder if my daughter will ever laugh at it with me. Maybe she will just think I am funny, and laugh at me, myself, instead. Serve me right.

Aunt Beatie. What did she think of all this? The thing was that I never levelled with her. I never said, "Aunt Beatie, I need to explain myself to you. Let's make a pot of Earl Grey and sit down and talk with one another the way we used to." Maybe if I had she would have known the answer? No, I don't think so. Outside her lexicon. Too healthy. I suppose my direction must have become clear to her as my law school years went by. I went to fewer party events. My Marxist papers from Europe and elsewhere lay in piles with their wrappers still on. I pleaded overwork. I did really have a lot of it to do, and that was always a valid reason for her; still I think she sensed what was happening. The day I came home from writing my last examination I met her at the door of the apartment—she was naturally on her way to a meeting—and she asked me how it had gone. "Fine," I said, "I think I aced it." "That's good, Hugh," she said. She looked at me very hard. "I hope you're going to use what you have learned in the defence of the poor and the oppressed. I hope so." And she went off; but not, I think, before she had observed that I could not meet her gaze. Actually I've become a bit of a specialist in insurance litigation.

When I was called to the Bar, of course she was there. And she came to my glossy new apartment sometimes for dinner, and admired my new furniture, and I made steak-and-kidney pie for her and we laughed and talked almost like old times. And of course she came to my wedding. Shortly after I had found Robyn (I went on a rather systematic search for a wife after I'd been with the firm for almost four years), I took her to meet Aunt Beatie. Robyn was horribly uncomfortable. Aunt Beatie still lived in the same third-floor walk-up in the same dingy building, and I don't think Robyn had been in that part of Toronto or in such a building once in her life. Aunt Beatie was fine. She had made Eccles cakes for tea. She looked tenderly at Robyn and joined with enthusiasm in discussing plans for the wedding. Robyn was telling her about the bridesmaids' dresses when the phone rang. We listened while Aunt Beatie delivered some terse comments on the line of the French CP in May '68 to a comrade who was preparing an educational on the student movement. She came back to the sofa all set to revert immediately to green sashes and ivory lace, but Robyn had some trouble adjusting herself. She is a very conventional woman which I suppose is largely why I chose her. She is the ultimate in presentability. That's why she made no argument when I started the custom of having Aunt Beatie for dinner with us regularly twice a month. After the meal I would take Aunt Beatie in here to the library, and pour her a Drambuie or the evening's Scotch. Then she would fill me in on all the news of the left. I was beginning to build up a collection of biographies and autobiographies of leading Marxists, and she liked to look at these and talk about them as well. I think Robyn was both bored and intimidated on these occasions, but since Aunt Beatie was my aunt and utterly respectable in dress and behaviour apart from her politics, she did not object. Once I heard her on the phone talking to a friend. "His aunt's here," she said, "you know, the one I told you about. The Real Radical." I felt anger and amusement at the same time.

So there was no final cleavage between us, no scene in which she told me what a craven lickspittle sellout I was. I think she loved me. I was family. And I think she had a profound belief that people do what they have to. Though she would say that was too mechanistic.

Aunt Beatie's dead. She died of a heart attack three weeks ago. She never looked ill. She kept her trim figure and upright posture and air of composed assurance into middle age. Her hair was greyer and her skin was drier and she wore her glasses all the time instead of most of the time; those were the only changes I saw. The doctor said, however, driving nails of pain into me with each cliché, that she was all used up and her heart was worn right out.

I bitterly regret that she died before my daughter was born. Perhaps I thought that the gift of a great-niece or -nephew would be a recompense I could make her for myself? Selfish again.

But one thing I can and will do. The name we had planned if the child was a girl was Kimberley Anne. I never liked Kimberley—I don't approve of the growing custom of giving children the names of places or things, like Chelsea and Flower—but Robyn is very fond of it. Anne is Robyn's mother's second name; it's all right by me. But one of those two names must go now.

Out with you, false Kimberley or pallid Anne. Make way for Beatrice.

Goodbye, Aunt Beatrice. It's all I can do.

On The Point

The lake in early morning is calm, brown-silver beneath mist which dissolves to vision's periphery as she turns her head.

So quiet.

Eight-year-old Louise is the only person; her footprint on the cool wet wood of the small bridge across the bay is the very first.

Each morning a new cobweb spreads in the angle of the handrail's last span. She ruptures the web, for Mrs. Ireton. The thread gums her fingernails, and she rubs her hand hard on her faded red shorts, each morning.

She has come from tangled green, thigh-high cold and wet, of the once-croquet lawn by the cottage. Up on the hill now, where the cows graze in the afternoon, the tubes of tough grass are bitten down to an inch. The stubble feels good underfoot, something between tickling and scratching. She looks down, to avoid cowpats. Warmth touches the left side of her thin body and goose-pimples prickle on the right. The sun will rise over the hill while she gets the milk from the farm, and on the way back she will be warm all over. She can tell that the day will be very hot. Top of the hill now, with the farm briefly in view and the cows coming slowly from the barn. Close by stand a crowd of giant pines and hemlocks, black-green and witchy. Down the hill, down down down through milkweed, Queen Anne's lace, clover, Indian paintbrush, goldenrod, purple vetch. Down to the gate, cold metal fatly beaded with silver, which opens with a harsh squinch-grind. This is the first noise of the morning.

Once through she is truly on the point, a long isosceles triangle of land which goes out and out into the lake. She has a glimpse of the beach. The transparent water moves slightly now, lips over the hard smooth sand, prepares to form waves. How good it would feel to step just there, where beach and lake connect. But the path curves away from the water and the mist is rising and the milk will be ready, and so into the little wood, birches and poplars. The path is narrow, hard, rooty, fringed with shining clusters of poison ivy; she must watch her step, but she knows where the best birches are and stops to look. Their leaves hang like earrings. The rising breeze moves them, and now there is clear sunlight on the fluttering green.

Up the little rise and out into the full new sun. Here lies the old flower garden. A quick look on the way past—an overgrown mass of a thousand shapes of green, with flashes of pink yellow mauve white scarlet purple orange blue like butterflies randomly throughout. A vapour rises, a dew-cooled fragrance. Now the real garden, bigger, vegetables in rows, but she hardly looks, for here, here is the farmhouse. The broad shallow steps curve down with time, and on the middle step there is a cat, always a cat, and the cat is always lumpy with kittens or gaunt from nursing them.

The screen door.

In the kitchen are two women, old and middle-aged, mother and daughter: Mrs. Ireton and her only child, Pearl.

□

Finally young Mr. Ireton knows, decides, says, that he must go for the doctor.

Maybe he just can't bear being with her any longer, can't bear her cries, tears, smells, can't bear the bloody helpless abandonment of her limbs. Maybe even if he goes out to the barn where he can't hear her, he can still see her? Maybe he feels sorry, guilty, angry at her for causing trouble.

He tells her he's going. Probably he waits for a gap in the pains, shoves his cold bristly red face down to the pillow.

The mare comes from the barn, snorting, flicking mane and tail at the snowflakes jumping all over her like summer flies. Now, now the young husband is frantic to go, but the sledge is piled with wood and he must spend whole minutes flinging pine logs into the snow. He's in at last, wraps the fur robe round. The mare starts down to the frozen gritty beach. At the edge of the ice, hoof raised, she hesitates. She always does, he knows that, but right now when he is doing what he should have done hours, days, weeks ago he just can't bear it and slashes the whip. Blood flies threadlike on to the snow. Again, and she moves to get away from hurting.

The trio of man and mare and sledge recedes across the lake. Smaller and smaller. Quieter and quieter. Now Mrs. Ireton, lying in labour in the bed upstairs, cannot hear anything except the gentle blowing of the windy snow—and presumably her own cries, if she is crying. Smaller and smaller, a tiny dark nucleus against the whiteness, spinning its filament of track. In dreams that filament is red, like an overlay in a medical text showing the highways of the blood. It begins to branch out, to form a frail scarlet network over the lake, over the snowladen trees, over the dark house—he can't leave an oil lamp burning for all those hours he will have to be away—over barn, fields, hills, a network frail, elusive, sticky, clinging everywhere.

Does Ireton look back? What would there be to see? A brown house on a thickly-treed point, a small brown house with a second floor gable. There is the cat at the window. From halfway across, the frozen lake and the bays have flattened into the shoreline, the off-shore islands have coagulated with the mainland; the house and the few shuttered cottages can't be seen at all. No, probably he doesn't look. He concentrates on his task. This quick invigorating drive through lightly blowing snow may even be enjoyable. He has more sounds to listen to than she; the mare's brisk metallic steps, the slap and jangle of the harness, the scraping runners of the sledge. Does he consciously listen? Does he really look at the mare's dull brown rump moving from side to side, at the clouds of her warm breath evanescing among the snowflakes?

Now he is past the big island. He turns south, down the inlet at whose uttermost tip, four miles away, huddles the little town where perhaps he will find the doctor. Just one look back now? There is only the grey-white frozen lake with the dark band of trees about its edge and the whitish hills rising beyond. No islands, no point. Is she there any longer?

□

Pearl gives her sweet broad smile to Louise and strokes her hair lightly. Pearl's own hair is curly grey; she is plump, soft, moves slowly. She walks silently to the long table at which Mrs. Ireton stands, rolling out pastry, and touches her arm. The thin old woman turns, peers sharply. She does not smile at Louise, but the child feels welcome and asks what kind of pie it is today. *Blueberry*, says Mrs. Ireton, and Pearl smiles again. While the daughter and mother go to the stone pantry for the milk, city child looks at the discs of cream-coloured pastry, the great bowl of dark blue fruit crusted with sugar and studded with dabs of butter. Six large pyrex pie dishes stand ready. This food is for the tourists who tent in the field beyond the barn. The child knows she is not a tourist, but one of the summer people; still she wishes that she could have some pie, just once. The linoleum is cool and slick underfoot. She sees a thin bar of fiery orange at the rim of the door of the big wood stove. The calendar by the pump shows a snowy field, a small stone house with a smoking chimney, a tiny lighted Christmas tree. There is a clean smell of eggs, milk, Oxydol, homemade bread. Also vinegar. Nothing here is like her home.

Old Mrs. Ireton gives Louise two glass bottles of milk, hung in a wire contraption which enables her to carry the burden in one hand. Pearl takes a long wooden spoon and dips into the bowl of fruit. Smiling, she puts the spoon to the child's lips. The sugary globes startle her taste buds, so cold, blue, sweet. Pearl chuckles as she opens the door for the child and watches her go out into the sunlight, past the gardens and into the little wood. The mother has already turned back to her pastry.

□

Mrs. Ireton must also wonder if he is really there. She cannot visualize his "there" exactly, as he can hers, but somewhere out on the lake, past the big island by now surely, perhaps even close to the far end of the inlet. Perhaps she does what many of us do when waiting through something horrible, interminable—*I will count to three hundred and then I will hear the sledge coming*, something like that.

Three hundred once, three hundred twice, three hundred thrice.

Or perhaps she does it by the contractions. *After three more I will surely hear the sledge.* But time in labour is not accountable. No point can be named. Certainly not in that labour, alone, the birth gone wrong, alone, the pain unbearable, and not a thing in the frozen world to do but bear it. Does she believe that it need not have been this way, if she, if he, had been different?

Perhaps she comes to that point at which we are sure that what we await will never come.

Someone (who?) will eventually find her, dead, with the dead child stuck like a little frog in the mouth of a bigger one. All the house will be cold as the stone pantry, the cat mewing savagely, the Christmas pies turned to leather on the long table in the kitchen.

A spider sits in her web in the middle of the ceiling. Of course it must be dead. They don't move in cold weather anyhow. But what if it isn't? What if it does? If it begins to reel out its line and waver downwards, floating, swaying, turning at first in the waves of her breath and then in the still air, down and down towards her, over her, around her? There will be no feeling. There will be snow a foot deep on the doorstep. The tracks of the sledge across the lake will be obliterated. That long purposeful line leading away from the point will be completely gone.

□

City child, city woman, Louise still doesn't really understand how the young Iretons planned to live, there, at the turn of the twentieth century. What was their vision? The country is now loud with cottages. But then—not wild, even then, close to empty, the Indians long gone. A half dozen summer places around the miles and miles of lake. A small farmhouse on the point. Some acres of stone-studded clay soil (city child makes snake-coils and squat bowls of the plastic dirt, dreams of staining them with brilliant vegetable and berry dyes, but they fall apart in an overnight rain). Some hillsides, thin coats of grass and bracken over rock, a thousand places for a cow to break her leg. A small stream (city child makes flowerboats, leafboats, knows a flat stone on which one day lie a dead frog and a white leech, next day half a frog and a fatter leech, etc.), a small thin stream which mushes out into swampiness which turns gradually into the lake.

The lake. In winter Ireton cuts ice and fills the icehouses of the summer people, stacks cords of wood by the cottages. In the spring, go round the lake and take off shutters, check pumps, repair docks buckled by ice; in the fall, reverse the film. He builds cottages, boathouses, outhouses, docks, sheds, as the land around the lake gets sold. (Not his land, yet.) He works alone. He's gone, dead, years before Louise's birth in the late 'thirties. That solitary labour is still mentioned first when his name comes up among the congregating summer people. Next: he was stingy, money-grubbing, mean. Next: he did beautiful work. With pride, the summer people show each other an Ireton dock, an Ireton stepway to the beach.

The barn. City child spends afternoons here, smelling the hay, chewing it, watching the spiders spin their webs, playing with kittens and calves. Cows. Gradually Louise learns that there was once an almost-scandal—the animals tubercular, the milk polluted, awful—all because Ireton wouldn't buy some equipment or have some tests done, something, something that would have cost money. That is another thing the summer people always mention.

The cows are all young Mrs. Ireton's work. Milking: milk cream butter; rinse, boil, fill, carry, empty, measure, count, and sell. Daily, like the hens. City child does not like hens much, but always goes round for eggs with old Mrs. Ireton, once, at the beginning of each summer. The summer she is eight they see the biggest cobweb ever made, a great floating wheel of silver filament with its creator motionless at the hub. Louise sees with amazement that a tiny swollen sac beneath the spider moves, moves tumultuously. She cries out. Mrs. Ireton's hand sweeps, slashes decisively down. Her foot stamps again and again.

The vegetable garden. Young Mrs. Ireton's work: sow, weed, thin, harvest, can, preserve, bottle, pickle, salt, and store.

The baking, the laundry, the housework.

The flower garden; obviously she found some time.

But how then did it happen? Why did she come to her first and only childbirth alone except for him, and then all alone? Money? Surely the doctor must have had patients who paid in chickens, in cutting wood; surely he would have come? Did no neighbour know that she was near her time? No one thought to drop by the house on the point and see if the young mother carried high or low, if it was her lower back that hurt or her swelling ankles? Were they all afraid to come, afraid of him? Did their husbands tell them, *No, stay out of it?* Did she ask him to get someone to come and did he say, *No?* Or did he just not answer when she spoke to him? Did she even speak?

Perhaps he thought she was making altogether too much fuss. Animals don't behave like this. Still—sometimes the farmer, the vet, the shepherd has to thrust hands or instruments into the choked birth canal to bring out the new creature. He must have known that. Perhaps he had even done it himself?

□

A deep unused verandah shields the front room of the farmhouse, and thus it is cool even in August. Each year, once, city child is allowed to enter. An old Axminster on the linoleum is extraordinary underfoot. There is plush on the carved couch. The lace curtains look soft, clingy, but are stiff with sugarwater. Painted flowers brighten silky cushions (city child tries it on a t-shirt and it all goes runny). Pot-pourri and furniture polish. Before she goes to sleep that night in her cabin with the water lapping the shore ten feet away, she tastes roses, spice, harsh lemon at the back of her throat.

Summer after summer, Louise looks at the sepia photographs in the oval wooden frames. The proud young man is bearded, moustached, has a strange high collar which looks uncomfortably tight. The young woman's dark hair is cloudily soft about her head, and a locket is half-visible in the lace at her throat. Who are these people? One day, one year, the old woman bends down to give her the milk bottles and a thread of gold loops down from the wrinkled pink neck. She sees the curve of the locket behind the faded stripes of the housedress.

But she was beautiful. . . .

□

Even if he had pulled calves and foals out of their slithery birthholes, trailing blood and amniotic fluid, even if he had dealt with placenta and feces and all the dripping mess of animal birth, perhaps he could not bear the doing of these things for her, by himself or by anyone else.

Or he was afraid that he would be too rough, too big. He would kill and that would be manslaughter, not just the unfortunate death of livestock.

Or perhaps he feared the mystery of her body.

Or infection.

Or perhaps he was just a skinflint bastard who didn't rate his wife as high as a mare or a cow, resented the fatigues of pregnancy which made her slow at her baking, cleaning, milking, washing, preserving, gardening, hated the look and feel of her great swollen body and wanted only for the birth to be over so he could get back to sticking it in her every night.

□

The silence of the farm kitchen draws the child. At the cottage, at Louise's home in the city, people talk all the time. There is a continual fluent counterpoint between her mother's German-flavoured soprano and her father's firm confident Canadian bass. Most often, at the cottage, there are guests, lots of guests, and a steady ripple of conversation in two or three or four languages amidst the scripts, films, wines, editorials, galleries, and novels. If there is no talk, there are said to be Gaps. (In her teens she will suffer greatly from Gaps.)

In silence, she watches Mrs. Ireton lift a big mound of soft dough, strips and flaps bobbling round its edges, on to the flour-sprinkled oilcoth. A few passes with the rolling-pin and the pastry flows out like an ironed sheet over a bed. Louise watches the muscles working in the thin old arms. Mrs. Ireton peels the knobbiest potato in one unbroken spiral. When she tops and tails beans they flip like tiny batons under her knife. The old woman bends over her work. The parting goes straight down the middle of her head with absolute precision. Louise is sure that if she could count the silky white strands springing up from the pink scalp, there would be exactly the same number on each side. At the back there is a perfectly round bun of hair, slightly yellowish here. Louise can see no hairpins. How does it stay up there? Mrs. Ireton is preparing the chickens. Ugly, bristly, flabby, dead. The long match hovers over breasts, legs, wings; there is a strange acrid smell; the wings tuck under and the onions go in, and the legs are bound. And there clearly is roast-meat-to-be, which never flapped and squawked around the yard. So quick, quiet, neat. The pink fingers of the old hands are thick, and do not flex far. The nails are always the same length, exactly. The child looks at her own brown grubby digits. She can stretch a full octave on the piano at home in the city. It is hard for her to cut the nails on her right hand, and sometimes her father exclaims at their length and helps her. She has learned this year a Schumann piece called "The Happy Farmer," at least that is the title given in her music book; but her mother, shocked, says No no, that is wrong, it is really "The Something Landmann, The Merry Peasant." What is a peasant?

Meanwhile Pearl slowly wipes the long table, the counters, with a white cloth pungent with Javex. She washes pots. She sets the tables for the tenters' dinner. Then she starts again with the cloth. Mrs. Ireton gets up quickly with a little *Tsk!* noise and shows her a pail of potatoes. Pearl smiles and slowly begins to scrub. The sun is all over the room. A bee's drone, falling and rising, falling and rising, inscribes itself on the screen door. Occasional small crackles come from the stove, and the rich smell of roasting chicken. Pearl pushes back the grey tendrils of her hair from her perspiring forehead, and mixed drops of sweat and potato-water fall on her plump arm, on the floor. Her mother (*Tsk!* again) gives her a hand towel and points. Louise, at last, finishes the little pan of beans Mrs. Ireton has given her, and the old woman tips the handful neatly onto the green pyramid in the waiting pot. The child stretches, and takes a kitten from the basket by the stove. Once on her lap the little creature purrs, licks itself, then turns its stomach to the sun and sleeps, its tongue sticking out. Pearl looks at child and kitten, and smiles. The forgotten potato lies in her damp lap. Her mother sets it in her hand again. Then Mrs. Ireton takes down the ironing-board, and everything that is crumpled and rough becomes smooth.

□

How long has Mrs. Ireton been lying there?

Long enough to feel great pain, to feel despair, to know the uselessness of tears, to reach quavering hands down under the heavy wedding-present quilts to the wetness between her legs and quick pull the red hands back again.

The cat, for a while, is on the bed with her. Is that some help, to feel the warm fur and hear the purring? But then it jumps down, so lazily, lightly, goes downstairs to forage and doesn't come back.

Does she scream, just to know she is there?

She can convince herself that the spider is moving.

What does she think of? How many times is she sure that at last she hears the runners of the sledge?

Pain and pain again. Winter darkness comes early, even though everything is white outside; he left in the early morning and in December they light lamps by one, two in the afternoon. There are no lamps lit, only the lights in her head and the silver filament above and the pain that comes and comes and comes, the baby inside her that isn't moving any more, can't move, is blocked and choked and caught in the place of birth.

But of course it is darker inside the house than out. When he emerges from behind the big island and turns for the homestretch across the stone-hard lake, the doctor beside him, the mare speeding up because she knows the barn is near and she is cold and tired and hurts where she has been slashed again and again . . . when he emerges, everything is pale grey, oncoming darkness suffused with snow. The sledge is much darker, a little blot or node of black, away, away across the lake, impossible to tell what it may be, only that it moves and is coming closer. Gradually the shape declares itself: horse, sledge, two people. Are they talking? What about?

Now the bays begin to open up to them, the shore no longer a plain black-green belt, and the point appears. The small brown house is shrouded in snowiness, no smoke. The fire has gone out as it had to.

Here in the home bay the snow is forming drifts. It is hard for the mare. The doctor (wearily? angrily?) loosens his fur robe and gathers his bag to him; there are metallic clicks within. The mare strikes sand beneath the snow. The runners scrape. That sound, finally, is heard. The doctor is out. He is running clumsily through knee-high whiteness to the farmhouse door.

Alone, Ireton goes to the mare's head and leads her to the barn.

□

Oxygen deprivation: the term and its implications are now commonly understood.

At the birth, something went wrong, was what Louise's mother said one summer when the young girl (not quite a child now) finally grasped that Pearl was odd in some way beyond acceptable variations of *normality*. The words were said neutrally; this was to protect; it was probably wise. What terms did the country people use when Pearl was a child, was growing up? Stupid, touched, not all there, dumb, simple, retarded, idiot, wanting? Something went wrong at the birth. How long was it before the mother, the father, understood? Did they talk about it? Did she say, *You almost killed me and look what you did to our daughter, our beautiful daughter?* What did he say?

There were eighteen years for such possible discussions. Then Ireton died of a stroke.

Once in a later summer Louise told her parents that Pearl made her feel the way the cows did. Her father laughed immoderately, spluttering over his espresso on the verandah after dinner; her mother's laughter was less, but still laughter. So she never said it again, but it was true. There were the same great gentle eyes, staring so calmly, the slow, leisurely movements, as though there were all the time in the world to dry a dish or eat a head of clover, the ample shape, warm, maternal; the sweet breathing smell. That same August, she and a visiting friend were climbing on the hills beyond the farm. The day blazed so they could almost hear the liquor in the trees, the bracken, the grasses being sucked out by the ardent heat, could almost see the leaves begin to crisp and wrinkle and flicker and fire. Down a little gully the girls went, and there was the skeleton. Huge hoops of grey bone. A strange angular structure with two enormous round holes in it. A litter of straight shards, imaginable inside legs. The bones were pitted like driftwood. Mastodon, dinosaur, woolly mammoth; but now imagination could not sustain these delightful fantasies. This was cow. They were frightened that if they looked too closely on the ground nearby they would find the tail. Somehow that would be the worst. So they ran hard down the terrible hill to the new municipally-provided gravel road, and went that way back to the cottage instead of going by the point. They jumped along the road to make great clouds of white dust rise and cover their tracks.

Nonetheless, it was Louise's father who, still another summer, told her much more of the truth. He came at the subject sideways, through complaints that no one now could do work around the cottage as well as Ireton had, all those years ago. And from there he moved into irritable recollection of how unpleasant Ireton had been—surly, taciturn, cheap. *Wouldn't even get a doctor when she had Pearl. Anyone could see that woman'd have trouble having a baby, she's narrow as a post. Oh no, wouldn't lay a cent on that. Probably scheming how to raise his prices while she was lying there. And finally he did go, and when the doctor got to her at last she was half-dead, torn to pieces, and Pearl like that. Probably held it up to her after that he'd had to pay doctor's bills. Twenty years old, she was.* He looked then at his daughter, fifteen, and frowned. Later that day Louise's mother sought her out. She talked well, most lovingly, but the calm of the sunlit kitchen was gone forever.

Forty years after Ireton's death, Pearl also took a stroke, as the country people there say, and died.

Louise, now in a distant city, heard of the death some months later, and that night dreamed of the scarlet filament for the first time. What had those four decades been like for Pearl? For her mother? In the name of God what were they like?

City child, now city woman, came with her firstborn to visit her parents, came to the lake to show her baby the paradise of her childhood. The little that was left of the Ireton farm had been sold, and Mrs. Ireton was about to be put in a Home. She was staying with a neighbour. She had shrunk; was quite deaf, and did not know who the visiting mother and baby were at all. She did not touch the little girl, but sat working her fingers together in her cotton lap. Louise longed for Pearl. There was conversation with the neighbour lady over Oreos and iced tea, and Louise then took Mrs. Ireton's knotted cold hand to say goodbye. The old woman looked up with that well-remembered peering glance, but there was no light in it.

Going back to her family's cottage, the new mother found that she could not bear to walk by the point, and so took the long way by the now-paved road, crying, and holding her healthy baby very close.

Evelyn and Rosie

Evelyn

I never worked so hard as in the war. Boeing, down by Coal Harbour. Roaring machines, outside salt smell, rain, mountains. Rosie and me, we'd leave work aching. A good tired, slept so well. We'd laugh so hard, us and the other girls, try not to snort when the foreman came by. We were doing important work and that's a fact, as Rosie would say. Felt good. Hardly a thought to the Germans on the receiving end. My dad was a socialist but I wouldn't listen. In the thirties Dad'd take Mum and me to those big socialist picnics at the Powell Street Grounds, potato salad and all sunny and the big voices overhead. Mind you, he wasn't the only one had a hard time getting clear on that war. The CCF, they went dillying around, the CP line was no strike. Some help, eh? Poor Mum. I'd get home, she'd be standing there tired, peeling some damn turnip, me all set to eat and run meet Rosie for the dance, and the zip'd go right out of me. "She doesn't do anything, Rosie! She's no reason to be tired!" Not cruel, you know. Just young. I didn't know the word depression, for people I mean. She wasn't nice to be around; I didn't stay. Then Dad'd be off to the Stanley Park Discussion Club.

Lovely silly music. Older people think the kids' music today's terrible. 'S'no worse than ours. It's your being young gets into it, forties or eighties. Oh, we danced. Rosie and me'd walk home then, talking and giggling, then up early for Boeing, feeling fine. Never time to say everything. I knew all about Italian Al; in the spring of 45 we knew it'd be over soon, and Rosie wrote him in France to say Yes, when he came home to Vancouver. She even had her kitchen planned, and she was going to lose ten pounds. How many times! The chestnuts were blooming. That's a lovely tree. On our street they were planted first a pink and then a white. I still go take a look every spring. Nice view of the water there too. So I told Rosie she was a fool and she said, "You'll be next."

The men came home, we all lost our jobs. Oh yes, we were upset, but we didn't see anything *wrong*, not truly. Engagements, weddings. I helped Rosie with her folks, Lutherans, they took on something terrible. I was maid of honour and Paul was at their reception. He was still in uniform. He could add three-digit numbers in his head; never seen anything like it. He couldn't dance, because he'd had his right leg amputated below the knee, so we sat out and talked. I talked. He's still in uniform in that picture from our engagement party six weeks later. I'm wearing my Boeing workclothes, see? For a joke, I said. Actually said. That thin dark smile of his. I couldn't look at that photo for years. Now they look like other people. So in June 46 I was pregnant, throwing up something awful. Rosie was due in July, and we sat in her kitchen and cried. Al wasn't coming home nights. I couldn't put a name to my tears. We believed we were just in a rough patch.

I wanted to call our daughter Rosie, but she's Brenda, like Brenda Frazier the Glamour Girl. Paul said ours would put that one in the shade. He'd never believe I didn't care about his leg. He got so tired, down and up all day on that milkwagon. I could tell when the not-there foot hurt. There's Brenda on the horse, that frilly white bonnet, and Paul holding the reins, grinning. Gives that milkcart a Wild West look, doesn't he? I had to take the picture so's his bad leg wouldn't be towards the camera. As if you could tell, under the uniform. But in 48 Fraser Valley Milk got motorized vans; Paul couldn't drive them. Angle of the pedal. Couldn't. Went round to Avalon then, the other small dairies, nothing. I've wondered—all those horses, for milkwagons, breadwagons, rag-and-bone carts, where'd they go?

Down to our last dime when Jimmy came in the spring of 50. New babies make you think. There's me, skilled worker, diapers and dishes. There's Paul, best of a leg in France, no skills but killing people and driving a milkhorse, weren't needed any more thank you. We were still at war. We are still. The other day on TV some general from the first war, they call it, he died, and all they said was he got the VC seventy years ago for killing seven German gunners single-handed. Is that a life? We weren't even generals. Now I've read, studied, but then all I knew was I had to find a way through. Paul was silent, you see. I said to Rosie,

"Why shouldn't I work and him stay at home?" She was Rosie, she saw I meant it, so she sat, thought. No kitchen curtains yet; Al's restaurant partnership wasn't going well, she hadn't told him she was pregnant with No. 3. "Evelyn, it ought to be okay. But it'd kill Paul if you suggested it." That was like a knock on the real door when you're dozing on the living-room couch.

Then a woman came by selling make-up. She had a poor approach, but right after I phoned her company. From day one I did well. The beauty was putting it over to Paul. Not a *real* job, not Boeing for sure, lots of women did it for pin-money. So I said he thought it was fine. I loved selling. That second when the door opens and you decide what line to take! A rush, like the kids say. Oh, I sold tons. Never wore it myself, still don't. Of course Rosie bought from me. She always wore hers vivid. To the end. How could Al take his eyes off her? Once he even made a pass at me. I never told Rosie. So while I was out Paul looked after the kids. He got casual work, doing the books for some neighbourhood businesses, helping people with income tax, never steady, scraps here scraps there.

Once Rosie and me took the kids on a day trip up Indian Arm. Beautiful day, the wind on the water, the mountains, and the boat had mesh railings so we didn't have to watch the kids every second. They looked swell together, the dark-haired and the strawberry blonds. Cold War then, I was bound we'd go on a Ban-The-Bomb march. Rosie wouldn't tell Al, of course. I'd asked Paul, but he wouldn't. "There's no use." He says so still. I have to *try*. That awful time with Rosie—she was forty-four, her youngest was seven, she couldn't bear it. I found the hack. Only the three of us ever knew. She almost didn't make it. Oh I can't think of it; but I found him, that's a fact. On the way back suddenly we had ourselves a thunderstorm, what a sight, rain rushing at the water and lightning zipper-zipper all over the sky. Jimmy was almost four, no not scared, he said, "Mummy, I hear drums, big drums in sky, what is?"

How hard he tried in school. Oh I heard it all, not achieving, uncooperative. He couldn't read properly. Paul couldn't bear it, was awful to Jimmy, Brenda wouldn't have mattered near so. I watched Jimmy's sweetness go under and that ugly don't-care come up. Come Grade Six he didn't do any better but he said Mr. Nilsson was okay. He was close to retirement, thick glasses and fishy eyes under when he took them off. He said, "Mrs. Harper, I've seen many boys like Jimmy. Not stupid, not bad, we know, don't we? Some little process in the eye, in the brain, is not quite right. There's talk now of putting men on the moon; someday we'll know why boys like Jimmy can't read." We both cried. Seeing the future. No big drums. And Al—when I see him with his "little lady" he married not six months after Rosie died, I'm tempted to tell him. But Rosie wouldn't want me to do that to her Al. We went on the Ban-The-Bomb march together, Rosie and me. First of many. Then I just went, I wasn't part of making them happen.

Rosie

Evelyn's always so pretty, her lovely curly auburn hair and blue eyes, and so slim, just a wand I tease her, and me such a dumpling, but *she's* never made me feel that. Such a sense of colour. You know redheads never used to wear pink? Well, right back in wartime Evelyn had this dress, russet and pale pink and dark pink, all blurry flowers, looked watery somehow, and was she beautiful. Brenda's her image, just fairer, and right away that's a terrible knot, the two so like and Evelyn and Paul hardly speaking, it all went for them years ago, and Brenda can do what Paul can, that mental arithmetic? A girl, and she can, they throw numbers around like kids do balls. It hurts Evelyn.

What did happen? Oh yes, we talked. More than lots of women back then, forties and fifties, and more than to the men, for sure. I'd feel embarrassed sometimes for Paul, "I know what you do in bed and you don't know I know," like that. For Evelyn and Paul, at the beginning, it was splendid. I'd had all my kids before I found out. Evelyn and me, we'd taken the kids up Indian Arm for the day, oh, a good time, gorgeous sun and a thunderstorm too, so happy, and when Al and I got to bed that night I went places I'd never gone before. He was jealous, see, wanted to get me back. Exciting. We made love as much as we could there for a while, funny, hiding from the kids. Oh, he was tomcatting again soon enough, but I knew now what she and Paul'd had—and they'd been new to each other. Evelyn had Jimmy on her lap in that thunderstorm, and the pair of them looked like at the circus. It's made a difference to my life, knowing. To Al's too, I think.

Now when my Al gets mad, he yells, stomps around, makes phone calls and yells—wears me out, but it's in the open. And he goes after what he wants. Upfront, like my Peter says. Paul takes the mad and goes down inside himself and buries it. Won't talk. Evelyn told me once he'd gone twenty-three days without speaking to her. I'd guess there were longer goes than that. So Paul was let go, the army, the dairy, and he never got back in, anywhere. And he never tried, really. Something like this: You don't want me, okay, I won't turn myself into something you *would* want. They got by. My Al got Paul work here and there but then not, said people didn't like working with him. He can come across really sullen. I think Paul tried to get at Evelyn and me over that. Then the company offered her a real office job, nine to five, supervisor of a sales team, a month's trial. She didn't tell Paul right off. Dumb, eh? I guess after an hour she could see a dozen things'd have her operation humming along in no time, must have been like fresh salt air to her, and like a fool she came home all enthusiastic and told him. "How could I do it, Rosie?" she said to me. "What did he say?" "Nothing. He just let me go on till I stopped."

She kept that job. Why? Rent, food, kids' clothes, that's why. I think she'd seen Paul wouldn't change. She used to imagine, see, imagine he'd take that crazy gift he had for figures and turn it into a paycheque. But we've gone over it fifty times—what if she'd chucked it, thrown herself on him and said, "You cope, I can't"? Would that have turned him? There's a dream she has sometimes, she's in a long hall and Paul's behind it, but her key's the wrong one or she can't find the door, and she goes up and down feeling the walls and can't get through. I said, "Evelyn, why doesn't he open to you?"

Good friends. How do people get through their lives without? Once Evelyn arranged an operation for me. I didn't think I'd make it through, and she held my hand. I guess you can see I'm not too well now, that's a fact, well, Evelyn's with me often, and we just sit and hold. Oh, of course, the children, but that's different. Now Evelyn, she had a bad bad time there with Jimmy's dyslexia. Brenda? No, always fine in school, especially math. Evelyn tells me Bren's going for computer work when her youngest's in school, not that she's even had it yet. Back then—always lots of girls over to play. Those make-up samples didn't hurt, I'm sure. Brenda never fooled with it like most, I remember my Lucy and Maria with purple rings on their cheeks, green eyebrows. No. Brenda knew what she wanted. We've talked about it, wondered if her ways came from the troubles at home. Once when Bren was eight she was bound she'd go to Amy Wu's birthday party; everyone knew white kids never went. Evelyn's got the photo still, two rows of Chinese girls, so solemn, black hair and straight bangs and dark dresses, and Brenda spang in the middle all curls and frills and smile. What I mean is, after that Brenda and Amy didn't chum around any more.

The last time me and Al and Evelyn and Paul really did anything together was trying to stop that apartment on the corner there, that Grandview Towers. Real sneaky, through the rezoning before the neighbourhood heard a thing. See, it blocks half the view to the water. People from Alberta. My Al got his lawyers on to it, but we were just too late and that's a fact. So Al said let's at least have a good dinner together, we'd go up to his Trattoria and he'd lay on the best. Paul said for the three of us to go ahead, he'd be along. We were having antipasto and Al was needling Evelyn, saying he'd seen her picture in the *Province* that morning, a peace demonstration, weren't the kids embarrassed? Really the people were just dots, but I wished he hadn't because Brenda was angry. Anyway. The way he looked at her I could tell; he'd made a pass at her some time and she'd told him to get lost. Right then Paul came in, looking so strange, and told us he'd got the caretaker's job at the Towers. Champagne. Evelyn got lit and she had that green dress with the v-neck and you'd swear she was about eighteen. Then this couple came to our table and Paul said, "Why it's Thora and John," and they were the Grandview Towers owners. John was maybe fifty and Thora not half that, pretty red curls and a witch of a smile, and I saw, and I saw Evelyn see, how things would be. That was 61, my Carlo'd just turned seven.

Evelyn and Rosie

Rosie said I should leave. Rosie. Al hadn't been faithful to her past half a year, she'd stayed, had the kids, held on, and now there were custom-made kitchen curtains, avocado appliances, a garburator. She just got fatter. Outsize dresses, bubbly ankles. Her kids loved her so. But Rosie—what she felt her life was about, she could be it in that marriage. Like a house with a lousy floor-plan that you've lived in for years. She knew I was different. Oh, I thought, all right. Maybe I got scared, or I thought about everybody else too much. Or maybe I'd already left. Whatever there was to leave.

Evelyn wasn't scared. Those anti-war marches in the early sixties, they weren't popular. People ganged up, yelled Commie, threw things. Once a cop brushed past her on his motorbike and she called right out, "You're just a uniform," and went right on walking. She did more and more with the Vietnam groups, every minute free from work and housework. Paul was mostly at the Towers. Al said Evelyn neglected Jimmy and Brenda; now they spent more time alone than my kids, sure, but Evelyn and Jimmy were clear anyway, that's always been sweet to see, he came to a lot of demos, and she and Brenda never were, so what's the diff? Brenda hated anti-war. Big fights. Mine with Lucy and Maria were all about boys, what time they got home, all that; Evelyn said she'd rather that, but even though Brenda was always dating she never fell for anyone.

Sometimes I felt I was my father with his socialist club *and* my mother all tired peeling carrots in the kitchen and Brenda slamming upstairs. Breakfast with the kids, office, meet, phone, plan, order; then the anti-war office, meet, phone, plan, decide—that's harder; home, supper with the kids and maybe Paul; then on to the phone or typing for the anti-war. Yes, all the factions; hard to steer clear. I just stuck to the Out Now line. Oh yes, tired. There'd be a spring action and a fall action and in between we didn't stop. But every day I'd talk with Rosie or see her. Teenage times. Pete and Maria quit church and I helped Rosie with Al; he took on something terrible. Brenda graduated in the sixties, yes. Jimmy'd dropped out before that. Finally he got work in an autobody shop, that was the first Nixon year, don't want to think about it. The guys there said he was a miracle worker. Nice, eh?

For Maria's grad I wore a silk I'd got at Christmas, and it was loose. Evelyn had me to the doctor fast! She decides quick. Those anti-war kids think she's terrific. She says it's like selling lipstick, you have to sense what people feel and zero in. Can't bear it that she couldn't do it with Paul, or if she could it wasn't any use. Thora, years and years. Once Evelyn said we had to go to a women's lib meeting, and they told about the women in China, Speaking Bitterness. Oh no, not that we didn't agree, just too close too late. We went on every demonstration for abortion rights, though.

Rosie used to tell Al it was dollar-forty-nine-day at The Bay. The dumb jerk never caught on. We paid a thousand to that hack.

I teased Evelyn I'd lose weight like I'd wanted to for years. No more chafing! Thinness carries clothes better, that's a fact. Paul now, in that green work coat he wore from the Towers. Evelyn was so angry at me she cried. I did too.

Rosie's neck reappeared and her waist, even her legs a bit, but all wrong, and the terrible skin colour. Then chemo, and her hair fell out. Her girls took her to be fitted for a wig. They were so tender with her. She still bought make-up. Al cried. I'll never forget that. Big burly guy. There was the Tet offensive. Brenda got engaged, Paul was so angry. I couldn't see why, Tim's nice, Brenda'd always planned for this. He shouted, "Remember what it was like for us? She ought to have that at least. She's not in love." Rosie never was in that place, I was in it and had to leave, which is worse?

I've seen Evelyn a grandmother, and glad of it. Love not mixed with anything. Jimmy first, he and Anne adopted a little girl. We figure some way he didn't want to father children himself. Paul's funny, talks about the baby in a different voice than about Brenda's boy. Not my Evelyn. She's a lover. She thinks of the world.

It dragged on. In 74 I found out the multinational that owned my company made napalm. Had for years. All mixed in with the Fiery Red and

Pearly Gold. Might as well have been at Boeing. Felt like twenty years a liar. I quit. Rosie helped me through. Just a stick in the bed she was then.

I hope Evelyn'll be all right. She takes things hard.

In 75 it all ended. Rosie died, the war more or less finished, John and Thora sold up and went back to Alberta. Paul was out of uniform again. It's the nuclear now. I work in the office; sometimes he comes with me and stuffs envelopes. After a long time, people are used to each other. When the Britannia Centre started we even went to some meetings to try and stop it. No use. Now there's the elementary and the library, the rink and the pool. People like it. I do myself; you can't do anything with a view. I talk to Rosie in my mind, tell her. Sometimes I can hear her voice.

Tabletalk

"Landmines. We are landmines. In Kelowna and here, waiting to blow." My neighbour at the dinner table was unknown to me. I saw a small dark head, the hair coiled at the back; long turquoise earrings, rimmed in gold, that trembled; a high-necked georgette blouse, paler turquoise, whose lines indicated a slender waist beneath. Her plastic conference tag was at such an angle that I could not read the words following "Hello! My name is."

She was looking at the nearby head table, where Mr. Andrews, the dean of instruction, was introducing the dinner speaker. Meanwhile the servers moved through the hotel ballroom, distributing appetizers in small footed dishes of thick glass. Three shrimps hid in a tangle of pale shredded lettuce. Mr. Andrews, incredibly, said that our speaker tonight needed no introduction from him. Then, consulting note cards, he listed Emil Zwiecki's academic achievements and contributions to distance education. This took some time. "International," he said at last, "an international man."

Many people clapped, myself included. Moving was a relief. My neighbour's hands were still. The hall's acoustics amplified the clapping noise and wafts of chilly air moved unpredictably about. Dr. Zwiecki, a stocky man who carried his rolled-up notes like a hammer, reached the microphone and began to adjust it to his height. Mr. Andrews immediately helped, and the two appeared briefly to wrestle for control of the metal stem, from which issued woofish sounds of apology and deprecation.

"This is intolerable," said the low voice to my right.

"Why did you come?" I asked.

The waitress, reproachful, removed our untouched appetizers. The other dishes on her tray were empty; the man on my other side had gobbled his shrimp and then swabbed the dish clean with a mop of lettuce impaled on his fork.

My neighbour looked at me as if assessing my intelligence. "Because I want the Institute to hire me as a telephone tutor again next term," she said. "Appearances are important." She laughed briefly. "Also, I have never had an aversion to free food." She turned to the head table again.

In his thanks to those who had invited him, Dr. Zwiecki moved on now from the names of specific individuals at the Institute to more abstract entities, such as a universal thirst for knowledge, posited as inherent in the species.

On her return to our table, the waitress set out eight bowls of clear brown fluid.

"Shall we?" I said. "My name is Helen."

"Myrna," said my neighbour. We lifted our spoons.

"And so," said Dr. Zwiecki, "tonight we consider the role of distance education in the community. Emerging in response to needs, it meets them—or it does not. The impact, what is it?"

"What is it, do you think?" I asked Myrna.

"Campell's consomme, with some oregano in it." She set down her spoon and sat back as if preparing to exercise patience. Her hair was not perfectly smooth after all, for tendrils and wisps lay about her temples. The ruffles on the front of her blouse had a childlike look. I judged her to be in her late thirties, like myself.

The soup was at least hot.

"How long do you think it will last?" Myrna asked, meeting my eyes for the first time. Hers were dark, lidded with turquoise, sad.

"A good hour for dinner and Zwiecki, allowing for thanks and so forth. Another for the business meeting after."

Myrna sighed and gazed again at the speaker. The waitress, removing the soup bowls, smiled at me. Zwiecki explained that distance education was available to all members of the community. "No one, not the most disadvantaged, the most marginal, is excluded. This truly is the open door education." The light glinted on the solid, sweating, middle-European face, the silver-rimmed glasses, the thin lips opening and closing. Cold fingers suddenly pinched my arm.

"Shall I tell you what is going on in my family, Helen?" asked Myrna in her low voice. "What has been going on, I should say. For over thirty years—but I can do it easily in an hour." Her look was intense. If I said No, would she rise shrieking from the festive board, shred her turquoise to tatters and stab herself with a table knife?

"Go ahead," I said.

A pair of waitresses arrived at our table carrying large trays loaded with dinner plates. These were evidently extremely hot, for the bare-handed women snatched them from the one surface and flung them on to the other. The roast potatoes slid about. There were two of these, small, on each plate, and something that resembled a wet brown washcloth, folded. This was beef. Between the protein and the starch lay six spears of canned white asparagus with a strip of pimento across them like an Order.

"I wonder when Dr. Zwiecki has his dinner," said Myrna. "I think the potatoes may be edible. This story begins with my mother and father."

"Don't they all?"

Myrna looked sidewise and smiled. Her eyes remained sad.

"They met in one of the camps, in the last days of the war. They were both almost dead." With her knife, Myrna guided the asparagus to the far side of her plate. The veins stood up on the back of her hand. Her nails were ridged, bevelled, perfect ovals of delicate pink. "Left-wingers they were, the pair of them."

Dr. Zwiecki noted that exceptional self-discipline was required of the distance education student.

"And you?" I asked.

"Oh no." She gave a little shake. "That's one infection they did not transmit." Having cut each potato into eight small chunks, she began to eat them *seriatim*. I attempted the beef. At the knife's touch it fell into strings and shreds, grey beneath the brown slick of gravy.

"So, like thousands of others, Jews, Catholics, who'd survived the camps, gypsies too for all I know, they came to North America after the war." Here Myrna's voice became a sing-song. "And they experienced great difficulties—ill-health, ignorance of the language, prejudice against foreigners, and they survived all those too, just as they had the camps. And they had children. My sister Rose in forty-seven and me in forty-eight." She chewed and swallowed. Her voice came some way back towards normal. "And from having crummy jobs, both of them, they got to a point where my mother didn't have to work. My father's salary was enough. And so they had won through. By that time of course we were almost grown up."

According to Dr. Zwiecki, the task was to prepare course materials which reinforced the student's motivation.

Myrna drove the tines of her fork through the slice of beef on her plate.

She then withdrew the implement and laid it down, aligned with the knife and the asparagus spears. "When Rose was eight, my father began molesting her."

Her voice sounded as though a contribution from the listener would be appropriate.

"How did you find out?"

"She told me."

"So you were able to help her do something about it?"

"She told me last week."

"This is 1985."

"Yes." Myrna looked at my nearly empty plate. "How can you eat it?" She looked at me, fully. "You're tall, of course, you can carry it. This molestation started very mildly. That is often the way. I happen to have read a good deal about it. She was always his little darling, his pet, you know, cuddling and kissing and bouncing on the lap, all that." Myrna spoke with extreme distaste. "There was a regular time, you see, when my mother and I were out of the house: the music lesson. How in the world they managed it I don't know. Violin for me, piano for Rose. The little Gerhardt girls and their music." Her voice was so dry it almost cracked. "So once a week he could get at her. The finger up the leg, you know? Under the little white panties, and in. The other hand up on the soft chest." The voice cracked.

A woman opposite Myrna and myself selected a roll from the platter in the middle of the table. We watched. To break the roll she had to use her thumbnail, as for detergent or cornflake boxes, and the food immediately disintegrated into dry shards. The woman, looking with concern at the heap, flicked perhaps half towards the platter. Then she gave up. Dr. Zwiecki said that the student needed to develop a sense of trust in the ability of the Institute to guide and assist him.

I could think of no reason to restrain the question. "Did he have intercourse with her?"

"No," said Myrna, regaining her voice rather loudly, so that one or two people looked at her, "no. Just the cuddling and the finger, and he came on a towel. You know, when she told me that I remembered my mother complaining long ago that my father was careless with towels, used too many. She had to do extra washes. I suppose that remained in my mind somehow. A small puzzle."

"Did Rose ever tell your mother?"

"Oh no. It went on till she left home, at nineteen. She got married."

Now that most of us had finished our main course, the waitress arrived with a bottle of British Columbia wine and poured for those who wished it a careful three-quarters of a glass. Myrna drank several times before continuing.

"No," she said. "Rose didn't tell our mother. In families like ours, you know, the children understand that the parents have been through dreadful times. Things were harder for them than for us here in Canada, too. We spoke English from the beginning, of course, with the kids on the block. My parents still have heavy accents, especially my mother. So Rose didn't want to distress her. She knew it would cause distress, she knew, even then. And on the other hand. Of course. She." Myrna paused, and drank. "She loved my father. She still does." Myrna spoke hurriedly now. "I do too. In many ways he's a lovable man. I'm sure she believed that if Papa wanted this it must be all right. It must be. It must be." She emptied her glass.

"And she didn't tell you?"

"No. I was the little sister."

Gradually Myrna's breathing calmed. The glasses at our table were emptying and conversation was becoming general; only by conscious listening could I hear Dr. Zwiecki explain that a relationship of special importance obtained between a student and his telephone tutor. The shrimp-gobbler tried cutting a roll with his knife. He achieved a slice, and the woman across the table looked envious, but when he took a bite buttery fragments sprayed all about him.

Myrna's left hand rested trembling on the table. She wore an excessive quantity of rings. I could not determine if any related to marriage.

"So why did Rose tell you, after thirty years?"

Myrna started. "Because our parents are getting divorced."

"After forty years?" I myself had divorced after sixteen, an experience akin to ripping off a limb without anaesthetic.

"Yes. They both want it. It's strange—what did you say your name was?—Helen. All the years of hardship, struggle, and now they're at the safe end, pensions, the house paid for. And they're finished." She raised her empty glass to her lips. "My mother wants to be alone. My father wants to visit friends, talk about the past, go back to Germany for a visit. The first ever. She has no interest in that."

I thought about this scenario. "What's your sister doing now?"

"She lives in Kelowna with her husband. Hans. He's German too. And their children. Little Hans, and John and Melanie. So—Rose told me because—it's as if—what did she hide it for, then? If they end up divorcing after all. For what the sacrifice?" Myrna's voice was speeding up.

I signalled for more wine.

"I'm sorry," said the waitress, not sounding so, "but that's all the free wine. If you want more you'll have to pay."

Our table consulted. The shrimp-gobbler resisted, arging that it was just like the last contract negotiations; we ended up subsidizing management. The woman across the way said that was all very well but we had a good two hours to go and she couldn't hack it without a drink. Consensus achieved, we ordered and obtained more wine. The waitress then abandoned us. Other tables had had their places cleared, but we still faced greasy remnants of beef and breathed the pissy smell of asparagus. "Rose told me over the phone," said Myrna, "it was awkward. I couldn't discuss it just then." She looked down at her hands. "I told her I'd be seeing Mama the next day, and she screamed down the phone that I mustn't tell her." She drank.

Dr. Zwiecki's voice—a little weary, I thought—was developing the theme of teamwork. I hoped for coffee soon. I wondered if there would be dessert, and what Rose looked like, and where her children and husband had been when she had screamed at her sister not to tell her thirty-year-old secret to their divorcing Mama. I felt stiff; the chairs were uncomfortable for a person of my height.

"Mama and I were in the garden," Myrna continued, "sitting under the plum tree. So pretty in blossom. She hates it. She can hardly wait to get into the little apartment she's buying with her half of the house money. No raking leaves, no picking up the windfalls, no jam and freezing." She drank. "We were talking about Rose and her family. Mama said, 'I don't like the way Hans looks at Melanie.' She looked at me. Old. Scared. And I thought of my own father and I thought of my niece and I felt my mouth opening and I told her."

"You told her about Rose?"

"No. About me."

Myrna held her wine glass with such a shaking hand that red drops flew out on to the tablecloth. They landed in a shape like a butterfly.

More and more rarely as life goes on do I act on impulse, but at this point I took Myrna's other hand in mine. She did not take it away.

Dr. Zwiecki said he could not emphasize sufficiently the need to welcome wholeheartedly the ever-expanding and enriching cornucopia of technological development occurring world-wide which could enable ever more sensitive response to the needs of the individual student.

All the other tables had dessert now. The waitress returned to ours and removed one or two dinner plates.

"He started on me the same way. The same age. The music lesson. Rose's was Tuesday, mine Friday. An hour and a half each. So he got his rocks off"—Myrna smiled, a brief rearrangement of her facial muscles, expressive of horror—"twice a week. It may be the national average. I'm not sure." Her voice was beginning to speed up again.

I thought about that average. It seemed far too high.

Dessert. In tall parfait glasses were layers of savagely green ice-cream and a dark green liquid. The little saucer on which each glass stood bore also two rectangular wafers, cross-hatched. I let go of Myrna's hand. She consumed her entire dessert without stopping, spoonful after spoonful of green sliding down under the turquoise georgette.

"You have no idea," she said, licking her long thin spoon, "what it is like to sit with your mother under her plum tree and know that the same man's penis has been in you both."

She was right.

"Oh yes, he did actually fuck me. He—led me to think. I understood. I believed. That he and my mother didn't do this any more and if he didn't do it with me something awful would happen. I didn't know what. I don't know what."

The shrimp man asked if he could have our wafers. Having collected them from all round the table (they tasted like toothpicks), he began to construct edifices along card-house lines.

"And you never told your mother either?"

"I tried to." Myrna looked past me towards the head table as did many others just then, for some modulation of Dr. Zwiecki's tone had signalled the approach of his conclusion. "The first time it happened. She wouldn't listen to me. She kept saying, in German you know, 'You're crazy, don't say such things.' And then she said in English, 'I have to work, we need the money.' And she kept showing me her hands and saying 'Look, they are all rough and ruined.'"

Dr. Zwiecki's lips opened in a broad smile. His voice shook a little as he said, "Especially for one who came to this continent as youth, in youth, after those terrible years in Europe, there is—believe me—great joy in working in education such as this."

The relief of clapping was enormous. The cold air rushed about the room, smelling of coffee. Soon the waitresses were circulating with large metal jugs.

"He was a teacher in Germany, before the camps," said Myrna, "and she worked in a department store. Lingerie. Of course none of that was possible here. She cleaned floors and bathrooms and he worked in a warehouse. The day she gave notice to her last employer they bought champagne. We had never seen such a thing, Rose and I."

Dr. Zwiecki relinquished the microphone to Mr. Andrews with jesting reluctance.

The cream at our table had grown a skin. Myrna pierced it with her spoon and smilingly passed the little flowered pitcher on.

Mr. Andrews made many announcements about the conference panels and plenaries scheduled for the next day. No one paid any attention.

Calmly, Myrna stirred her coffee. "He understood a lot, my father. When I wanted to go to university, leave the town where we lived, change my name."

"When you married?"

"Oh no, I mean my first name. Minna. Mother couldn't see why I wanted to go to the city, but he sided with me, helped me, insisted that they give me money."

"So do you think your mother truly didn't know?"

"She'd certainly hidden it very far down. When I told her the other day she was eating a date square and she stopped chewing in the middle of a mouthful and went pale. I've never seen her do that before. Then she took the half-chewed stuff and threw it into the bushes and said, 'So it was true. God forgive me. Don't tell Rose.'"

The waitress arrived. Clashing the parfait glasses together by twos and threes, she slid them on to her metal tray and then distributed dishes of mints. The shrimp man asked if there was any more coffee. She smiled broadly and said "No."

Myrna offered a dish to me and took a mint herself. "On the weekend I'm driving up to see my parents again. My sister and her family are coming too. It's to be a family reunion, before they divorce."

The mints, unpleasantly sweet and acid, tasted like soap.

"Do you suppose—"

"I don't suppose for a moment that my father thinks anyone knows anything."

"Will you—"

Myrna stood up. Similar movement was beginning all over the room, for people were determined to stretch and socialize before the evening's agenda overcame them. Picking up each of the mint dishes in turn, Myrna threw their contents all over the table. Most of the candies fell to the parquet floor, scattering and bouncing like pebbles in an explosion. I rose also.

"What has caused me so much trouble in my adult life is that quite often I enjoyed it," Myrna said. "Goodbye."

As she turned away, I saw that the turquoise georgette was not a blouse but a dress, which moved gracefully about her as she walked up to Mr. Andrews. Smilingly he introduced her to the honoured guest. Dr. Zwiecki had just got his dinner, and even at a distance I could tell that the gravy on his plate had congealed.

A Young Girl-Typist Ran To Smolny:[1] Notes For A Film

Street corner in Burnaby.[2] Hot summer day. Time now. Three young women. One's cute, one's punkoid, and one's neither. She looks as though she's trying to be someone.[3] Some of her clothes fit. Punkoid smokes and Cute giggles. All three carry sheaves of a revolutionary newspaper,[4] subscription forms, pencils. Street: small tidy stucco houses, small tidy green lawns; a few larger, scruffier frame dwellings; some humans at work with hedge clippers; gnomes and flamingos stand about; curtains[5] drawn everywhere.

Cute, giggling, "Isn't the anarchist house around here?"

Experienced Punkoid frowns. "Yeah somewhere," indifferently.

Unsure one has millisecond vision of black dwarves massing by the barbecues, TV aerials flaming with black flags.[6]

Dissolve to Punkoid's cross face and sharp voice. "Kate, you take this

---

[1]So says A. J. P. Taylor in his Introduction to John Reed's *Ten Days That Shook The World*, Penguin, 1982, p. xv. NB see if Trotsky or Deutscher refer to her.

[2]Boring bedroom suburb of Vancouver, British Columbia, Canada.

[3]Lots of room here for directional creativity. Wd. be interesting to see what F and M directors would do. (In the women's dorm at college in the late 50s the phone-message book included a column showing initially the sex of each caller. Still done?)

[4]NB devise suitable name.

[5]Priscillas.

[6]Consider music for this bit—in fact consider soundtrack throughout. Shd. be highly patterned. Maybe silences for all the inner visions though?

block, both sides. Meet you back here in an hour. And *don't* get tied up talking, okay?'' Cute goes one way, combing her hair, and Punkoid another, not. CU on Kate, whose adam's-apple[7] jigs as she swallows; not attractive.

The stairs of the first house are cartoonly tall, looming. Kate can tell there's someone behind the glass-windowed and curtained door.[8] Her knock sounds explosive. East Indian woman in sari, festooned with small children, opens.

Kate, brightly, ''Good morning! I'm visiting in your neighbourhood to introduce our[9] paper, *Da Da Da Da*[10] [EIW remains expressionless—camera on her throughout Kate's remarks], and I'm sure it will be of interest to you. We cover the international scene from a revolutionary perspective,[11] and each monthly issue also contains [here Kate's voice starts running down like an old Victrola; camera is still on EIW's unchanging face] up-to-date analysis of events in the Canadian labour movement, the women's movement, the. . . .'' Behind the EIW appears a handsome teenage boy. He takes a paper from the top of Kate's pile, looks at it, hands it back, says ''No.''[12] His glance moves over the little bump Kate's right nipple[13] makes in the close-fitting fabric of her shirt, and he closes the door in her face. Immediately Punjabi bursts out multiply within.

Kate walks farther down the street.[14]

Next house. Flowers planted geometrically. Razor edge between grass and sidewalk.[15] Door, freshly-painted, bears sign in stick-on letters. The

---

[7]Obvious question, but let's ask it anyway: What about eve's-apple? If it's supposed to be all her fault, how come she doesn't even get the damn thing named after her? Or did the apple just slide down her evil throat as smooth as milk?

[8]Shot here through letter-slot, about eight inches from bottom of door—the sari and the bundle of kids' legs.

[9]Note that this is never explained.

[10]Making up a name will be hard. Sarcasm and flippancy are easy, but irrelevant here. Can't use any extant. Many of the non-extant have too many connotations, resonances (go easy on that lit crit stuff, okay?).

[11]The extreme disjuncture between the contents of these two sentences is echoed in the changing tones of Kate's voice. Readers/viewers may feel the impulse to laugh; they are invited, since they are so smart, to think up other and better ways to do door-to-door selling of a revolutionary newspaper in late 20th century North America.

[12]He does all this quickly, decisively, no hesitation whatsoever.

[13]The left is concealed by the revolutionary press.

[14]This sequence can repeat exactly whenever needed. Kate has to look extremely alone. Camera behind her, the long street stretching out.

[15]Could have shot of a man using an edger here.

Jone's. Kate winces. Then we see her from the back as she positions herself at an angle to the doorway so as not to seem invasive to the opener. She knocks gently. Middle-aged man, prim-faced, appears. We see Kate's lips moving tentatively and then with more determination, and come in on "women's movement, the anti-imperialist and third-world struggles. We[16] are also active in the NDP across Canada. . . ." Man disappears. Victrola effect. Kate waits, nervous. Man reappears. Kate's eyes are wide open and eager, the man's narrowed and suspicious.

"We take the *Trib*."[17] He closes his door, precisely and not noisily.

Kate walks farther down the street.

Next house. Geometry and razor-edge again. Kate pauses at first step, arranges sheaf of papers, has pencil ready. We watch her from the top of the stairs this time; she comes up sturdily and rings the bell in a confident manner, stands feet apart directly in front of the door. Woman opens main door (screen door remains closed): early forties, tall, glasses, curly grey hair, big build. Stares. We see Kate's lips moving again, and come in on "across Canada. Right now we're offering a special rate, twelve issues for five dollars and three back issues thrown in free. [The woman stares with a little smile. Kate is unnerved but continues.] Or if you like you can buy a single issue now and see if you like it.[18] [The woman's smile widens.] I'm sure you will, you know. It's really a very good paper. . . ." Victrola effect again, and Kate shifts her feet about miserably.

The woman says, "I'll take a subscription." Kate fumbles delightedly among her papers. There is a mixup over who will open the screen door[19] and whether Kate will give the form or the woman will simply take it off the top of the pile. Kate watches while the woman deliberately, too slowly, fills in the form. There is something not quite right about her

---

[16]Never explained.

[17]Newspaper of Communist Party in British Columbia. Kate's reference to the New Democratic Party tipped her prospect off that her newspaper is likely Trotskyist. (Though he would say "Trotskyite." Why this curious distinction obtains could be the theme of a very interesting little dissertation. Cf. Marxian, Marxist; economism, economist; racist, racialist; socialite, socialist. Consideration in order of distinctions between/among American, Canadian, and British usage.)

[18]Mistake. At the training session back at the hall Kate was told to suggest a single-issue sale only if *sure* the prospect wouldn't take a sub.

[19]A metal screen door, ugly, noisy, with stiff latch and handle. This model, now ubiquitous, is in every way inferior to the wood-framed screen doors of yore, the sound of whose closing on summer days has been eulogized by innumerable middle-class writers doing an E. B. White "Once More To The Lake" number.

manner.[20] Kate connects with this fact at the moment when the woman hands back the subscription form. Her accepting hand hesitates, indicating her realization that this is not a "real" sub.[21] But it *is* a sub. Kate leaves. The woman watches her go down the steps to the sidewalk and turn to continue her journey.

Kate walks farther down the street.

She thinks about that young girl-typist who ran to Smolny, ran with a message of crucial importance that Trotsky had to receive or . . . ? What kind of place was Smolny, anyway? A suburb like Burnaby? Kate visualizes herself running from Georgia and Granville out to Boundary and Hastings. But 1917, in Russia . . . fragmentary images of snow-howling steppes, Lenin addressing the crowd, onion spires, Peter and Paul.[22] Kate shakes her head, annoyed at her own ignorance and romanticism, waves her arm to brush the images away. She says, "I can't see how to see it." And the young woman herself, what is she like? Kate puts her on a country road, straight and flat, files of elms to either side,[23] and dresses her *seriatim* in professional Ukrainian peasant dance costume, Canadian Girls In Training uniform circa 1914, and 1920s flapper style. In all cases she has long blonde hair coiled in braids about her head, and she runs with long powerful strides.[24] Kate shakes her head again.

Next house. Frame, three-storey, beat-up but handsome still. Children's

---

[20]The woman has a look something like when a person who ordinarily wears glasses goes without them—drugged, lost, out of synch. Her clothing also can suggest mental illness (not the garments themselves but the way they are worn, the movement of the body beneath them).

[21]"Real" would be when the prospect and Kate had engaged in a brisk, vigorous, informed political discussion *as a result of which* the p. was convinced (not persuaded but convinced) to buy the newspaper. Kate does not believe she is able to carry on such a discussion. For the present she is right.

[22]Also possibly the pram-on-the-Odessa-Steps scene from *Potemkin* and the massed pennanted spears from *Alexander Nevsky*.

[23]A Dutch landscape, Hobbema or Ruisdael.

[24]Kate is thinking of the Modern Library logo, and then by association of the Everyman Library motto, "Everyman, I will go with thee and be thy guide/In thy most need to go by thy side." Everywoman is not mentioned, though many do use books as armour. For example, the woman to whom Kate has just sold a subscription: The day she ended her marriage, she went to the Kingsgate Branch of the Vancouver Public Library and read Dick Francis's *Rat Race* in its entirety. With that padding of 150-odd pages of cheap thriller between herself and the morning's experience, she felt able to go home and tell her children.

There are limits though. In 1975 gifted British Columbia poet Pat Lowther was hammered to death by her unadjectived husband in the bedroom of the East Vancouver home they shared (you could say). In her most need, where were you, Erato?

toys, tricycles, playpen littering front yard. Stacks of beer cartons and old newspapers on front porch. Door with long oval window, *no* curtains. Kate is still abstracted when she rings the bell; she waits, thinking. The runner image again, now in modern Olympics outfit complete with torch. Rings again. Door opens. Young man, plain, intelligent, scruffy.

"Yeah," he says, "don't I know you?"

"At the demo last Saturday I think," she says. Her face is shimmering between friendliness and urgency to begin her spiel. Young man's eyes drop to newspapers she's carrying. Face changes,[25] lengthens, goes tight, works in an ugly way. Mouth opens.

"Kronstadt!" he bellows. "*Kronstadt!* Trotsky murdered them!"[26] Glare. Spit.[27] Slam.[28] Glass of window in door vibrates.

Kate walks farther down the street, her step uncertain. She turns; she has only done four houses; she turns and the long street stretches out.

"Well. That was the anarchist house."

She walks.

"I wonder if anyone knows what her name was?"[29]

Now a sequence in which we don't see or hear the details of Kate's experiences. She goes up and down steps, from house to next house.

---

[25]This must happen *instantly*.

[26]The sailors of the Kronstadt naval base on the Baltic were among the most loyally revolutionary supporters of the Bolsheviks throughout the revolution, and Trotsky was a particular hero of theirs. In 1921 this relationship soured; the Kronstadt navy "mutinied against the Soviet government, and took possession of the fortress and two ironclads. After a bombardment lasting many days, the Soviet troops made a night attack across the ice and the revolt was crushed with much severity." So says the *Encyclopedia Britannica*, 14th ed., 1929. (Incidentally the excellent entry on V. I. Lenin in this edition is by L. D. Trotsky.) The *Britannica* then austerely concludes its remarks on Kronstadt by noting that the port "is icebound for 140 to 160 days each year, from about the beginning of December until April," and does not comment on Trotsky's approbation of the Bolshevik decision to crush the sailors' revolt. Whether this action should or should not have been carried out is a question which has since generated savage and unhealable differences among leftists.

[27]He's not had time to work up a good supply of saliva; nonetheless a visible glob lands on Kate's shirt.

[28]This is a really purposeful slam.

[29]If Trotsky knew, he didn't tell; he identifies her as a "working-girl from the Bolshevik printing-plant" (*History of the Russian Revolution*, Sphere Books, London, 1967, Vol. 3, p. 194). Was she a typesetter, not a typist, in fact? He also says she was accompanied by "A worker" (male, obviously), and that these two "ran panting to Smolny." NB remember to see what Deutscher says if anything.

A couple of times the pencil and the subscription form pass back and forth, there is briefly a smile on her face, her step is resilient—but essentially what we watch is the strengthening power of routine exerting its influence upon an individual.[30] Kate is passing through her "first time on the sub-drive" initiation. She is learning how to do this work of the revolution. It *is* work, she now realizes, combining in itself work's too-common attributes of difficulty and tedium. Occasionally there are brief flickering images of the young girl-typist.

Kate finishes one side of the block and crosses the street, and this new side stretches out and out before her. She is just half-done, is already tiring. She starts down the block. At the third house the camera meets her on the porch, a quick approach, nothing special about the building. She knocks. Brief wait. Then heavy sounds beyond the door; it opens. Tall, big-chested old man[31] with crutch under his right arm, newspaper under his left. His jeans and check shirt are clean and neat but he is not a person who is interested in them. He looks intently at Kate, who speaks without much expression.

"I'm selling *Voice of Revolution*, a monthly newspaper. We cover [the man blinks at the copy Kate extends to him] the international scene from a revolutionary perspective [his eyes widen], and each issue also contains up-to-date analysis of events in the Canadian labour movement, the women's movement [frown of puzzlement], the anti-imperialist and Third World struggles. We are also active in the NDP [his lips frame the initials CCF] across Canada. Right now we're offering a special rate [a smile begins to form on his cheeks and he looks Kate up and down; she becomes self-conscious but is determined to get to the end], twelve issues for five dollars and three back issues thrown in free. Would you like to examine[32] our new issue for a few moments? I'm sure. . . ." No Victrola effect this time; the man's deep warm voice simply obliterates hers.

"The *Voice*! Are you selling the *Voice* then?[33] Come in, yes, do come

---

[30]This is a tricky concept in revolutionary work. Too little routine in an organization leads to sloppiness and inefficiency. Too much generates an inability to respond to anything unfamiliar. An organization afflicted by this particular type of blindness is said to be suffering from routinism.

[31]Camera can explore this last adjective. Conclusion: this human is not as old as he initially appears. Poverty and hunger do a fine job in achieving this older-than-you-are look. So do the heavier emotions. So sometimes do intellectual struggles.

[32]This should be "look at," but at least she's saying the right thing this time.

[33]This telltale sentence construction is the kind that encourages right-wing Letter-to-the-Editor writers in deploring the too-strong influence of Scottish trade

in, put it down here so I can take a good look. The *Voice* now! I've not seen it in years."

They go past a small fanatically clean living-room and ditto kitchen and enter a TV room currently metamorphosed to a guest-room. The base is clear but the superstructure comprises open suitcases, books and magazines flung about, a couple of canes, a tobacco tin, some Guinness bottles. A crocheted throw on the hide-a-bed has been tossed back as by one arising from a nap.

"Now I'm Mac, Mac Ferguson, and what's your name, comrade?"

"I'm Kate Steele." Her face is nervous (inside house of totally strange man, remember that look on the porch), excited (this must be an old member, what luck), shy (what can middle-class Kate possibly have to say to this towering hulk of a worker?).

"Been in long?"

"Just a few months. I joined in the spring, after all the cutbacks demonstrations the teachers had."

"Ah, an intellectual. That's good, the party always needs some. Think of Trotsky himself."[34]

Abruptly Mac sits down on a chair and feels for his glasses. Kate sees them lying on top of the TV and hands them over; he does not say Thank you, just puts them on and is at once absorbed. Kate looks amused.[35] Mac reads, muttering. "This NDP convention now, what's the line?[36] They'll never go for that, the social democrats . . . That's right, tell the leadership where to get off . . . And this nuclear . . . terrible thing, terrible it is. Well . . . never win a strike that way, never happened that way yet on God's green earth. [Kate sits down on the hide-a-bed.] Latin America. Poor damn buggers. Poor damn buggers. Here there's food anyhow . . . Abortion rights. I don't like that now, no no . . . [Kate's face sinks; is she going to have to argue that one with this old comrade?] The International. The Four Eye . . . Ah, the masthead." He reads

---

unionists in Our Canadian Labour Movement.

[34]Kate does not consider herself an intellectual and is bothered, but does not know what to say.

[35]Such a reaction would be suspect in many left-wing and feminist circles. I like her amusement because it shows she has relaxed a bit and is focussed on him, not herself; likely it's a relief after her experience of the last while *not* to be the object of attention.

[36]This term, meaning the official opinion or analysis of a revolutionary organization, has been subject to much abuse in the North American left from (a) people too cowardly to accept the political commitment which *having* a line demands, and (b) people unable to move a political inch without the starch of an organizational viewpoint in their spinal columns.

in intense searching silence, then drops the paper and looks at Kate. Fragmentary vision of the young girl-typist, lights of a town ahead of her now along the narrow road, with darkness coming. "Nobody I know left."

Kate takes the leap delightedly. "Were you a member then?"

"Aye, yes, we're comrades, eh? Not really though. I dropped to sympathizer—oh, years ago now." He reaches for a tobacco tin, wincing; quite often in the rest of this scene he shifts his right leg about, seeking comfort. He starts rolling a cigarette. Kate watches his hands. CU to her face; adam's-apple activity again. Then she says, "What happened?"

Silence while Mac finishes rolling his cigarette, lighting it, dousing the match, inhaling, looking at Kate. Quite a long silence, which must establish that there is going to be an answer, that it will be serious and truthful, and that Kate is going to hear things she hasn't heard before.

"Comrade,[37] I honestly don't know if you could understand it without being in the Cold War. I mean, I started with the movement in the Depression, the Left Opposition, I went through the War—how old are you?"

"Twenty-seven," says poor Kate, feeling it a most puny age.

"The Second War. Oh yes, terrible, but the Cold War . . . that McCarthyism . . . what that did to people on the Left. Oh yes, here in Canada too, don't ever let them fool you on that, it didn't stop at the border. Awful divisions. In the movement. We were responding to the outside pressure, you see, it was hard to know that then. Fighting for our lives, we were. Oh, the factions. Comrades dropping out and dropping out, just a few of us . . . when you're so very few you're precious, see? You love each other and you hate each other too, somehow. Things go bitterly wrong."[38]

"Was it for personal reasons that you left then, Mac?"[39]

"Hard to say.[40] And I was a marked man in the union too, see. They got at me every way they could, the right-wingers, and they had plenty.

---

[37]Camera on Kate's face through most of this speech, which comes slowly, with pauses and gaps.

[38]Now to Mac's face, looking at Kate. Shows (a) the gratitude of a deeply lonely person talking to someone who at least understands the terms he's using; (b) consciousness of her youth and his age; (c) consciousness of her sex.

[39]Kate thinks these would be separate from and inferior to any political reasons, no matter how misguided the latter. She thinks so in spite of a year's participation in a women's group and a careful reading of Rowbotham's *Beyond The Fragments* and analogous works. She doesn't even know that she thinks so. Consideration of power, depth, and subtlety of bourgeois ideology is in order.

[40]Mac is not interested in answering the question but in saying what he wants

Fight to be allowed to speak at meetings. Hell, fight to be told the meetings were happening! Fight to get work. Fight to keep it. And to watch those "labour statesman" trading up their cars every two years . . . boom, see. Oh it was a bad time. My wife left me[41] [this very quick, en passant]. And then I got this," touching his right leg. "At Powell River. The mill. They wouldn't even take on the Compensation board like they should have. I don't get half the compo I should, not half." Mac is going to develop this theme, but recalls himself with a glance at Kate. Brief image of the girl-typist runner, freeze-frame, as if listening.

"So it was very bad. This was the early sixties by now, I was still trying to get back to work with this leg, took me another five years to realize I'd never work in the mill again. I was still trying. And we were having a big faction fight in the movement. Was right round when the NDP was forming, eh? No more CCF. Again and once again, the question of the NDP.[42] So at this one branch meeting I got up to speak, and the comrade in the chair ruled me out of order. And—what's your name again?"

"Kate."

"And Kate, you know what I did?" He looks in appeal at her.[43] "I cried. Couldn't help it. It was like everything was all of a piece, the movement, the union, my wife, the right-wingers everywhere. I cried. [Kate sees the girl-typist, clad now in a fifties-style skirt-and-sweater set, running along narrow city streets lined with old European apartment-buildings. She must be tired but she keeps a steady pace. She does not carry any parcel or envelope, so the crucial message for Trotsky must be in her braided head.[44]] And I walked out then, and I never went

---

to say anyway. This is true of about ninety percent of all answerers to all questioners.

[41]Throughout Mac's story, speculation is encouraged *re* whether he would have spoken thus, or even considered speaking thus, if the revolutionary sales-person arriving on his doorstep had been male. Interested readers/viewers may care also to imagine versions in which inhabitant of house and sales-person are respectively female and male, or both female. Such time-wasting fantasies can be most instructive.

[42]Kate has heard this sentence spoken with sarcasm and as invective. Mac says the words with plain sadness.

[43]This expression contrasts with Mac's big masculine head and face in a way commonly thought more moving than if it appears on the face of a woman or a child.

[44]Deutscher tell us that on 23 October Kerensky banned *Rabochyi Put* (title of *Pravda* since the July days) and ordered its editorial offices and printing-press

back." [CU on Mac's hands, which are large and thick-fingered; the tops of three fingers on the left hand aren't there. The ropy veins on hands, wrists, arms stand out in high relief. Camera shifts to Kate's face, which shows that she cannot think of a single thing to say.]

"Now you'll be wanting your money," Mac says suddenly and heaves himself up. "My daughter, she keeps me short, but I know where her bus-money stash is. No no, none of that, the press has to pay its own way, eh?"

Mac and Kate go into the kitchen's antiseptic neatness and he gets a cocoa tin down from a cupboard and counts out twenty quarters.

"Och yes she'll fuss, but what can she do, see? Next month I'm off to the one in Williams Lake and she's a lot easier on me—so I'll put her address on your form." He does, in a clear spiky hand which owes nothing to H. B. McLean.[45]

"Do you live part of the time with one and part with the other, then?"

"That's right. When I'm not in some damn hospital with this [slams leg]. Neither of them wants me, really, but what are daughters for, eh?" He folds the receipt and gets out his wallet. "I'm useful still, though. Do repairs here, look after the kids there. Her man here, though, he's an awful right-winger. 'Trade unions are not political organizations' [this in an assumed whiny tone]. Thinks it's terrible when the young Turks now,[46] they bring resolutions on Nicaragua to the meeting. I tell him

---

closed. "A working girl [NB he puts her *first*] and a man from the press rushed to the Military Revolutionary Committee, saying that they were prepared to break the seals on the premises of *Rabochyi Put* and to go on producing the paper if the Committee gave them an effective military escort." (Isaac Deutscher, *The Prophet Armed: Trotsky 1879-1921*, New York/London: Oxford University Press, 1954, p. 307.) Deutscher further states that this proposition, "breathlessly made by an unknown working girl, comes to Trotsky like a flash."

[45]System of handwriting instruction favoured for many years in Canadian public schools, and designed to pulverize any individuality in students' methods of moving pens across paper. Each lesson begins with "Ready for printing—Desks cleared—printing materials ready (practice paper, pencils, and compendiums on desk). Pupils adopt attitude of attention . . . All pupils should sit in a comfortable, hygienic position." (H. B. MacLean and Grace Vollet, *The MacLean Method of Writing*, rev. ed., Agincourt, Ontario, Gage Ltd., 1966-67, *passim*.)

[46]The OED says that this term identifies the Ottomans who in the early years of this century tried to rejuvenate and Europeanize the Turkish empire. It also defines the expression as applicable to anyone having qualities "attributed to the Turks," i.e. unmanageability and violence. By quoting a question posed in a 1908 newspaper, "Will the glorification of the 'Young Turk' kill this expression as one of reproach to be used in the nursery?", the OED suggets that the child-classifying expression used in Britain predated the 20th c. political term of reference.

he should go down Stateside and help Mr. Reagan." Mac and Kate laugh briefly. Mac takes a photograph from his wallet. "Here I was."

A young soldier, smiling. A casual snapshot. There is nothing to say. [The young girl-typist has reached a large building, windows lit from top to bottom, guards' figures black against the grey of the coming night.[47]]

"Will you take tea? Och, I should have asked before. No, you've to get on, I see that."

Kate is out on the front porch. She is down the steps and almost at the sidewalk. She looks down at her own arms and legs.

She sees the young woman now mounting the steps of the big building, Smolny, about to step inside tall ornate doors, about to break the seals of history and vanish nameless. The doors swing open and the typist turns to show Kate's face, an old face, worn, heavy, heavy, with the weight of human years. Then she's gone.[48]

"Kate! *Kate!*" Cute and Punkoid are down at the next corner, waiting and waving.

Kate walks farther down the street.

She turns in at the next house. She will complete her assignment.

---

[47]By one of those ironies Clio frequently produces—so sharp that in any "creative" or "fictional" work they would be roundly condemned as vulgar—Smolny, which was a building in Moscow, had in its previous incarnation been a finishing school for bourgeois young ladies.

[48]Trotsky, according to Deutscher, had been waiting for some provocative action from Kerensky, and the seals were precisely that. He sent riflemen and sappers to guard the printing-press, the machinery rolled once more, and the next day the rising started. (Deutscher, op. cit., pp. 307 ff.)

Presumably if this specific provocation had not come to Trotsky's attention some other would have, the revolution would still have taken place, etc. However, it did take place in just this particular way because of the thought and action of one particular young woman. For further discussion, see History, role of individual in. Plekhanov's essay is as good a place to start as any.

To Make a Good Death

"Less tonic another time," said Rebecca.

Andrew took the old woman's glass from her hand, went over to the sideboard and poured more gin. He stirred the drink so bubbles soared from the bottom and sheathed the lemon slice in iridescence. Adding a fresh bendable straw, he returned to Rebecca's bedside and bowed to her.

"Madam."

She sucked. "That's more like it."

Andrew inspected his aunt, lying in the high hospital bed they'd set up in the dining-room.

"You're down too low. I'm going to heave you."

"No I'm not. The bloody nurse can do it when she barges in next time."

His aunt's shoulder-blades felt like windshield wipers as Andrew slid his arm beneath her. At the back of her neck, the vertebrae were high white knobs under the bedjacket. The whole shrunken body moved lightly up the bed at a gentle tug.

"For heaven's sake stop mauling me and sit down where I can see you. Pull that red footstool closer."

"I've never like that footstool, Bec," said Andrew, moving a dining-chair. "I always sat on it when I was little and we came to see Grandpa, and I hated the feel of the plush on my legs."

"Well." Sucking her drink, she focussed her large grey eyes on him.

"That's why he made you sit on it, of course."

Andrew looked past Rebecca to the open window at the end of the dining-room. Rhododendrons and lilac and mock orange glistened from the afternoon's rain, but only the smells of medicine and old sick human were inside the room. He pictured that distant boy.

"I don't know, Bec," he said, and sipped his own drink. "Children get these ideas. I thought I had to sit there."

Rebecca closed her eyes in the way that meant pain. Andrew checked his watch; no morphine for a long time yet.

"Same with Deirdre," he went on. "She thought she had to have the light on in the hall outside her room, at night. She couldn't see the moon and the stars properly with it on."

"Who made her think she had to?"

Andrew sighed. "I don't know, Bec. It's at her mother's house. At my place she sleeps on the foam in my room and we both look at the sky. I had a big skylight put in."

Rebecca opened her eyes and glared. Andrew met her gaze, smiling.

"Humph," she said, and looked down at her diminishing drink.

They were silent, companionable.

Behind the stiff red brocade, the sheer white curtains lifted and fell in the evening air, revealing again and again the shining flowered green, and each time the old mirror above the sideboard shimmered, lucent in its mahogany frame. The dining-table, reduced to its minimum, stood to one side; next to it were ranged its many gleaming leaves; the big chairs with their wine-red seats clustered about awkwardly. Andrew thought of Christmas dinners, Thanksgiving dinners, grandparental birthday dinners eaten here, the silver, the big table napkins, the cranberry glass pitcher used for caramel sauce.

"All the meals I've eaten in this room," said Rebecca. She moved uncomfortably on the white-draped bed. "I wouldn't make much of a dinner now, would I? Old boiling fowl. Pretty bony." They both laughed.

"How's your drink, Bec?"

"My drink's all right. Why d'you let Deirdre stay with that woman?"

Andrew's hand, out to pick up his glass, stopped.

"Bec. 'That woman' is Deirdre's mother."

"I know that. She doesn't do her any good. I worry about her when she stays there. You just said yourself—trying to make the child scared of the dark."

Andrew finished his drink. He held the glass up-ended over his mouth until the cold of the ice cubes came through his upper lip to his front teeth. He judged that his voice could now be level.

"I've said this to you before, Bec." The old woman turned her head away. "Deirdre loves both her parents." The words were stiff on his tongue. "When Sheila and I split we agreed to honour that, to share her upbringing. So she spends equal time with each of us."

"Rubbish. A child needs reliability. Not this switching round all the time like a teetotum."

Andrew smiled at her. "Does a teetotum really switch? What is it, anyway?"

Rebecca's face turned to him again, the grey orbs focussing on the query.

"Get the dictionary. No—don't. Don't try to get me off the subject, Andrew, I won't have it. She ought to be with you. Sheila doesn't deserve her. You know it as well as I do. Worthless woman."

Leaning over his aunt, Andrew took her glass and put it on the little rosewood table by the bed. He clasped her hand of white bones and high blue veins; his hands were those of a man in healthy middle age, big and warm.

"Rebecca. Aunt Rebecca. She *is* with me, a lot. And we're happy when she is. Surely you can see I'm more content than in years? Better this way than a child caught between two reliably at war, Bec. And she loves Sheila." Yes, she does. "Would you have me tell a child not to love? Believe me, dear, it's all right. Now I'll get us both another."

Making the drinks at the sideboard, Andrew looked tenderly at his aunt in the mirror. Her profile, reduced to its essentials, was a hard outline on the pillow, but elegant still. The eyebrow angled up. The ear was sadly bare, for she said the jewels felt too heavy. As a child he had loved to watch Aunt Bec's earrings sway; always, she wore long drops of Coast jade. Deirdre wanted to get her ears pierced. Her mother was agreeable, but he didn't like the idea somehow. Twelve, so young. What had Sheila said that Deirdre said? "I have this idea of myself, of how I'd look."

When he gave Rebecca her second drink, tears stood in her eyes.

"Aunt Bec, what's this?" He stroked her cheek.

"And you don't even have proper furniture because you're giving that woman so much money. Sleeping on the floor." She sobbed, exhausted.

"Here now, here now, we'll stop all this. You rest, and I'll read to you. Where's the Wodehouse?" He got the book from the pile on the table and took some time finding a particular page. " 'The Fiery Wooing of Mordred.' How about that one?"

"Oh, it's a fine one." Rebecca's voice shook, but she was wiping her eyes and her look was happy. "There's that wonderful part about his rivals and their names, remember? Go on now, start reading."

As Andrew's voice rolled over the familiar sentences, he thought of the time when Bec had first read them to him. Fourteen, down with measles, he had lain in dimly-lit boredom for days of wasted summer. Then came his vigorous sharp-spoken aunt with books called *Right Ho, Jeeves* and *The Best of Wodehouse*. "You might be a little young for this, we'll see." They both laughed so hard Rebecca frequently had to leave off reading. Day after day she came and read to him, the green earrings trembling with her laughter.

" 'Mordred read this communication six times in a minute and a quarter and then seventeen times rather more slowly in order to savour any *nuance* of it that he might have overlooked.' " Smiling, Andrew looked up to share the line, but his aunt's eyes were closed. The gin tilted in her hand.

"I want another blanket," she murmered, "I can't get warm all through. Tell Kay to have the soup hot. I'll sleep till dinner." Andrew removed the glass. Immediately sleep took Rebecca. Her chest rose each time after a noticeable hesitation. Andrew checked; the bed was thick with coverings.

He went over to the window. The warmth of the summer evening had dried off the shrubs and grass and striped the dark greens with chartreuse. He thought of Deirdre leaping in and out of light, of Aunt Bec dying behind him. Why was she so fussed about the child and Sheila? He ached at the thought of Deirdre with her mother, the two of them, happy. The ice clinked in his glass. A summer sound, that, the cottage by the lake. Grandpa read aloud there, Dickens, Thackeray, pinesmell and lapping water behind the deep voice. The fire hissed. Once in that measles summer Andrew had told Aunt Bec he thought she and Grandpa were both terrific readers. She had immediately stopped, got up, left him. He had inquired of his father, Rebecca's much younger brother.

"They've never got on. He's—not easy, Andrew. Father's much gentler to you than he ever was to us." Andrew had frowned. His father had asked then, "Haven't you noticed that she never stays in the same room with him unless she has to?"

Andrew had observed his relatives. When Grandpa came into the living-room after dinner, his aunt busied herself in the kitchen with Kay, or made phone calls, or went to her study. At family outings to theatre or concert hall she sat as far from her father as possible. Andrew never saw her initiate conversation with him.

A starling flew across the lawn, changing from grey to black to grey. Andrew turned to look down the long room at his sleeping aunt. Such a lover as she was, why had she never married? On the plain side of handsome. Intelligent. Acerbic. Bookish. Still . . . had no one ever been drawn by the intensity of those eyes? From the kitchen came smells of meat, herbs. Surely father and daughter had spoken sometimes? All those years alone, together, in this house. Twenty-five at least, after Grandma died. Andrew shook. Father and son—well. Dad was soft, amiable.

Andrew finished his drink, and set the glass on the sill so Aunt Bec could notice it and scold him.

If he could convince her of Deirdre's joy.

What was the expression? To make a good death. There was something she wanted before she went.

The bird settled on the lilac and began to preen.

"I want my dinner." Rebecca's voice was cracked from sleep, fretful. Then the nurse came in without knocking, a known irritant, and Andrew left the room, for although he felt bad for his aunt and what she must now endure he knew that his watching would only make it harder for her; she was of a generation for whom the presence of others was not support but invasion.

Returning, he found her sitting erect, stiff with effort. Her cheeks were so fallen in they looked as if someone had taken an ice-cream scoop to them.

"Everyone abandons me," she said in a voice thick with rage. The door opened and there was Kay with the dinner trolley. Rebecca smiled beautifully at Andrew, anticipation brightening her eyes.

Soup. Tortellini in rich scalding fragrant chicken broth.

"Oh. That's good," Rebecca breathed, slurping in the steaming fluid from the old soup spoon that looked like a shovel in her wasted hand. "So hot. So good." She ate and ate. Andrew looked at Kay. And indeed when Aunt Bec had finished the plate of soup she looked without interest at her delicate portions of veal, grilled tomato, wild rice. "No. I don't want that, Kay, take it away."

Andrew ate all of his excellent dinner, aware that his aunt took pleasure in his enjoyment. Pale rose bloomed in her cheeks. The resurgence might last an hour. Having wheeled the trolley out to the kitchen, he returned with coffee and home-made petits fours. Rebecca poked at the pink fondant.

"What's this? I don't want any."

"Yes you do. Kay overheard you talking about these the other day, so she made them."

Rebecca frowned. "Mother used to make them. Then when I turned twelve Father said I couldn't have them any more, my skin would only get worse." She made a fist and struck the coverlet. Her face went suddenly from fierce to gentle. "Did she really? I'll try."

Andrew ate four of the achingly sweet little cakes while Aunt Bec took small bites at one. "You finish it for me, Andrew. And take a couple for Deirdre. I'll tell Kay how good they were. Now. Where's the pack?"

Lighting her cigarette, Andrew said, "No inhaling, remember."

"Oh I won't dear. Or just a little bit." Rebecca blew out smoke, and comfortably rested the hand with the cigarette on the bedcovers. Andrew contemplated her.

"You know, Bec, for a person with chronic emphysema and severe arthritis and terminal cancer you look remarkably contented."

She grinned, skullish. "I have my moments."

"Shall I read more Mordred?" Andrew waited patiently for a response, smoking, watching the soft dusk come into the room. The mirror was dark as a mountain pool.

"I won't be long now, Andrew," said Aunt Bec, her voice lighter, higher than usual. Perhaps she had spoken thus as a girl? He saw smoke pulling into her mouth and bent sharply towards her. Smiling, she exhaled. "Wretch of an interfering nephew!"

More silence, peaceful, relaxed. Gauloise smoke drifted in layered streaks towards the window, broke up in torn strands of veiling.

"I've never cried for my father, you know," came Rebecca's voice again, "not at the funeral, and not since." Concerned, Andrew leaned forward again, but his aunt's face was calm and her body at ease. The smoke rose from her cigarette. "I don't want to get into it all. There isn't the time. But he never laid a finger on your father or me. He didn't have to. It was all inside."

She inhaled. Andrew watched her chest expand, hold, fall, watched the smoke emerge smoothly, and sighed in relief. She looked at him sideways, smiling. Her expression became reflective once more.

"All inside," she repeated. "I'd thought my life would change when he died. Oh, of course in lots of ways it did—the release! But. Not deep down. Not me. I'd had . . . I'd had an idea of what I could be. And I just couldn't be it, Andrew. I'd used up too much of me withstanding him."

Andrew thought of his father, who had not withstood. In the dimness he could feel his aunt marshalling her story.

"When he was dying I was away," she began, speaking in a lower tone now, "doing graduate work in Toronto. At the library school. When I came back." Pause. "A walking skeleton, Andrew, walking towards me down that hall out there and calling 'Becky! My Becky!' He shook like a marionette." Pause. "He hugged me so I felt his ribs under the dressing-gown. It was dark red wool." She grimaced. "Later he asked me to help him up to his room. We got as far as the third step on the staircase. He sat down and cried."

After a moment, Rebecca closed her eyes. She looked like a corpse, smoking. Andrew waited.

"Then he began to speak. I couldn't make out what he was talking about. Then I realized. Poetry. Chesterton, awful mawkish stuff." Her lips moved and then her voice came again, and Andrew had to lean closer because occasionally she did not reach a terminal syllable.

> "My friends, we will not go again or ape an ancient rage,
> Or stretch the folly of our youth to be the shame of age,
> But walk with clearer eyes and ears this path that wandereth,
> And see undrugged in evening light this decent hall of death;
> For there is good news yet to hear and fine things to be seen,
> Before we go to Paradise by way of Kensal Green."

Andrew thought perhaps she had fallen asleep, although the cigarette was firmly upright between her fingers. Then her eyes opened and she looked sharply at him.

"I wouldn't have believed he'd have memorized junk like that. And when he finished, he sobbed and sobbed. I didn't know what to do, Andrew. Here was the monster of my life, crying and dying."

"What did you do, Aunt Bec?"

"I thought, 'What would I do if this were someone I didn't know? If it were an old man on a park bench whom I'd never seen before?' And I thought, 'I'd probably sit down by him and put my arm round his shoulders.' And I thought then, 'Can I bring myself to do that?' And I did."

Rebecca inhaled. Then she gave the stub to Andrew and weakly blew out. He tucked her chill hand beneath the covers and watched the grey fatigue encroach on her face; the skin went even thinner over the cheekbones, the hollows deepened till shadow filled them. Deirdre's cheeks were round, fresh, blooming with freckles, like Sheila's.

"I had a dress then," whispered Rebecca. Was she finally wandering? "It was divided. Dark green on one side, pale the other. Very smart. I could never wear it again, Andrew. Oh, such terrible feelings."

Andrew got up and walked to the window, where the shrouds of the curtains swelled and sank, and where the scent of flowers, released by the night air, poured now into the room and blended with the French tobacco. Outside was dark green darkness.

"Andrew? Have you gone away?"

He was by the bed, turning on the lamp, taking her clutching hand.

"You mustn't let that happen with Deirdre, with her mother, you mustn't, Andrew. She'll be Deirdre of the sorrows, she will. Those feelings kill, Andrew. The child, the child is what matters."

The eyes glittered with anxious pain. Andrew looked right into them.

"Aunt Bec, Deirdre wants to get her ears pierced. She says she has an idea of herself, of how she'll look. Sheila's giving her the piercing for her birthday."

"Twelve's too young, Andrew!"

"No. Her mother's right. The idea. She understands." Andrew's voice was full in the words.

His aunt's hand fell from his. "Give her my green ones, then." She sighed deeply and began dreadful coughs. Andrew flung back the blankets, put his hand firmly over Rebecca's diaphragm, pressed down. There was a convulsive movement of the entire body. Eventually it stopped. He stayed thus for a few moments, making sure, and then replaced the covers, tucking her hands well under; they were now very cold.

Rebecca breathed.

Andrew kissed her cheek and went with wet lips to summon the killer of pain.

Roses Are Red

"Our birth is but a sleep and a forgetting." So wrote Wordsworth; so quoted Queen Elizabeth, admiringly, at the opening of a modern institute for training in obstetrics. Some things I can't forget happened during my labour with our first child, but finally the doctors did more or less say, "Oh let's knock her out and be done with it," and did. When I came up through nausea and confusion I was doubtful about the red wriggling creature they brought to me and said was Colin. Was he mine? Stories of mix-ups in name bracelets. Unlikely in a small hospital in a smallish Maritime city, but I could believe it. Of course Roger was not with me during the birth; this was before such a notion was at all acceptable.

We moved then, to head office in Toronto, and Janet was born in a large urban hospital. Roger was away that time, in England. I was frightened, of course, on my way to the hospital, but felt also that my body was working in a more confident manner than the first time. The contractions came regularly and bearably. However, my labour had begun late at night. I became very tired; they said I could not push effectively and gave me an epidural. In the morning, though, I saw my daughter emerge from me, knew her immediately as mine, and was glad, for she was the last child we planned.

Then we moved again, here to Vancouver. Roger was promoted to responsibility for sales to the Pacific Rim countries, and travels now in the Orient a great deal. Our shared life continues in its I suppose rather peculiar pattern. The children and I are together all the time, and I am happy with them on the whole. I read and read and read. Getting Janet and Colin settled in kindergarten and Grade Two in a new school and city, and fixing up our new house, seem to take most of my energies. So I read, and watch TV, and put up wallpaper. By the time Roger returns from a trip I am spilling over with material I need to express and get reactions to, but that is not so easy for a couple after an extended separation, when there are many other practical and tangible things which cannot be delayed it seems. I think perhaps they could, but I am so warmed and invigorated by his talkative energetic presence, the children enjoy the expeditions we go on so much, and I so relish hearing what he has done and seen that I don't declare myself very strongly. I suppose that is what happens. Then after he has left again I think of all the things I wish I had said.

Recently I found myself pregnant again. I have read enough to know that this does not happen to educated middle-class women in their thirties, whose older children are already in school, for no good reason. At some level then did I will this? Did I deliberately put too little jelly on the diaphragm, insert it carelessly? Not that I remember. Roger is delighted at the prospect of a third child. He loves Colin and Janet extravagantly. I feel remarkably little. Oh, predictably there is fatigue at the thought of the night feedings and the diapers, and joy, and curiosity too—our son and daughter are very different in character, and I wonder what this new child will be like—but none of these feelings is very strong.

The one thing I am clear on is that I want this birth to be different. This will definitely be the last, for I shall have a tubal ligation after the birth. When I enter that world for the last time, I want to know what is happening, and why, and how. I decided to enrol in a series of childbirth classes. I told Roger, and he said, "But darling, why do you want to do this? You've had two already, what could you possibly learn?" He was to leave for a two-month trip for exactly the period that the classes would cover. But I went ahead, and booked a babysitter, and on the first night of class I went out of the house feeling almost excited.

When I entered the room in the church basement, I saw ten heavily-pregnant women lying in a circle on ten pillowed slabs of green foam. Behind them on spindly metal chairs sat ten men. On the floor in the middle of the circle were piled books and pamphlets on childbirth, breast-feeding, nutrition. I found myself a place and lay down. Some of the couples were talking to one another and I felt that the others were looking at me, alone. I wanted to pick up something and read but made myself not. I must not give in so easily. Clearly the women had all dressed themselves with as much care as I had for the occasion. It was strange how the vivid smocks, the shining hair and tinted lips seemed to draw attention to the lump each carried before her. Most of the men wore double-knit leisure suits in pastel colours. Roger did not wear things like that and again I felt out of place. The men sat shifty-eyed, trying to size each other up without being noticed. One had a roll of bristly fat oozing over his collar.

The teacher came in. I was surprised, I think everyone was. I suppose I had expected a childbirth teacher to be soft, maternal, charming? But this was a tall bone of a woman, with a long buck-toothed face, big hands and feet, intelligent grey eyes. I liked her at once. Her stomach under her neat denims seemed positively concave. I saw Boar-Neck's face give a little twitch of distaste, which I think she felt, for she looked nervous as she sat down with us on the floor. I sensed that she was going to begin with a joke, was anxious.

"Well, good evening, everyone," she said too rapidly, "My name's Peg. That's Peg, mind you, not Preg!" A nice clear English voice. People laughed, too readily. She took a deep breath and said more slowly, "Now I imagine that most of you women are happy to be here, and that you're looking forward to learning as much as you can in these nine weeks of classes. And I imagine most of you fellows had to be dragged here and wish to goodness you were someplace else. Right?" She knew exactly what she was talking about, I could tell from her tone. The couples looked at each other, smiled, looked at their teacher again.

Then a mocking male voice with an English accent said, "Very skillfully done, Peg," and I along with everyone else turned to look at the handsome man who had spoken. How had I missed him in my look-round? He wore a good three-piece suit, like Roger's, and while the other men sat with feet twisted round chair legs he leaned back as if in a La-Z-Boy recliner, displaying his long frame and legs and expensive gleaming shoes. His smile said, "And how are you going to cope with *me?*"

"Thanks," said Peg straightforwardly, and met his stare. After a few seconds he chuckled and broke the eye-contact. Peg looked relieved. She asked us to introduce ourselves.

The man who had spoken leaned down and whispered to his wife, a pale young woman with long floating hair who wore a plain blue linen dress. She looked a lot younger than he, and then I realized another reason why I felt so out of place—I was the oldest woman there by years. No wonder some of them had looked at me so oddly. The young woman's face did not change when her husband spoke to her and took her hand. When he sat upright again she removed her hand from his and placed it protectively over the enormous hump of her stomach.

The voices continued quietly round the circle. I'm Eleanor. I'm Don. This is our first child. This is our second, I'm Shelly, he's Martin. We have an adopted little boy and I'm so happy to be pregnant. (Smiles of empathy.) I had a hard time with our first, the doctors were terrible (puzzled frowns, Peg's among them), and I'm hoping this one will be easier if I can really learn from this course. Sandra, Wayne. The doctor thinks I may have trouble carrying to term. Now me.

"I'm Linda. My husband's away on a long business trip, so that's why I'm here alone." But would he have come with me anyway? Would he have stayed, after seeing these people who were not our kind?

"Don't worry," said Peg comfortably, "you'll learn it all and then teach him when he comes home. Is this your first?"

"Oh no, my third. I didn't—well, the first two experiences weren't that good."

"That's what we're here for, to make it better."

The lovely young woman now.

"I'm Grace."

"I'm Garth," rather loudly. "This is our first. Not for want of trying, though!" And he gave a big wink at the man next him. Boar-Neck and several others recoiled. We all looked at the floor, at the walls, at Peg. She looked intently at the next couple in the circle.

"I'm Jane," said the woman nervously. Peg smiled at her, Jane's voice picked up strength, and the process continued till all the couples had introduced themselves. I looked at Grace. The stillness of her face had not altered. She gazed down at her hands on the baby-lump. Some of the other women were also looking at her, puzzled.

Then Peg began her introductory talk. As her clear friendly voice traced the history of natural childbirth practices from Dick-Read to Lamaze, Kitzinger to Leboyer, I felt myself relax and become absorbed. This was what I had come for. The attention of the class was total. Some people asked questions. Peg encouraged women who had had babies before to explain or comment by describing their own labours. They were not very articulate, but I knew what they were talking about, for the power came clearly through their words, the immense taking-over power of that experience. Grace listened and listened, her hands rubbing back and forth slowly over her baby. I did not speak, though Peg's glance invited me to. Garth stared continually at Peg and she avoided looking at him.

Then we had a juice-or-coffee break, and the smokers were sent out into a little ante-room. Grace reached for a book and began reading attentively. I lay and watched the class. Garth joined the smokers. At first the group looked sideways at him and did not move. He began to talk to one of the couples. I could see him turning on the charm, I've seen Roger do the same thing at parties, and saw the woman laugh. A moment later there was a sort of ripple in the group, and there was Garth in the admiring middle. One of the men looked curiously at Grace, who was turning a page. What would Roger make of this? I had a letter half-written to him at home; how would I describe this class to him? Peg was checking through some file-cards. Once when there was a big laugh at something Garth had said she looked up. She seemed nonplussed and returned to her work.

After the break Peg began teaching us how to do the first level of special breathing, for use at the beginning of contractions. A-level: in nose, out mouth, in nose, out mouth, very gentle blow-out, barely bend a candle-flame. All of us women lay flat, eyes focused on chosen points of gaze, breathing quietly, absorbed. At first I felt silly but that passed very quickly. The men were doing it too, and the ones I could see didn't even look embarrassed any more. There was such a good feeling of learning in the room. Peg came round to check that each of us was doing the breathing properly. She came last to Grace and Garth. Grace was staring at the coatrack as though seeing a vision. Her hands still moved rhythmically over her stomach and I could hear her smooth regular breaths.

I could hear Garth doing the breathing too, with a kind of satiric perfection, as if saying, "Dearie, this is just *too* easy for someone as clever as I am." Peg sat back on her heels and gave him that straightforward grey gaze again. He snorted with laughter and stopped. Peg said, "That's fine," as if to a serious student, as if trying to convert this sarcasm to honest effort. Nothing on Grace's face. Peg got up; she was ungainly, I saw contempt flicker in Garth's expression, I saw her flush.

He said, looking up at her body, "Tell me, is this all just theory for you? Or do you have experience in the field?"

"Yes," she said simply, looking down at him. "All instructors for this course have borne children. My husband and I have two." He began to smile and she turned to the class as a whole. "Any questions about what we've done tonight? Okay then, practise the pelvic rock and the A-level breathing, and I'll see you next week." She began picking up the books from the floor. Garth glanced around and several of the men grinned at him. The women looked blank. Then they heaved themselves up from the pads and the men sorted out coats in a jovial bustling way. There was a lot of cheerful talking in the parking lot.

On the way home I realized that the thing between Garth and Peg had to do—partly anyway—with language. I have learned from Roger about the gradations of accent in England, and I knew enough to recognize that Garth's manner of speech was higher-level than Peg's. His and Roger's were alike. I looked a long time at the blank half-page of my letter to him. Finally I wrote, "The first class in the childbirth course was very interesting. I'll tell you more about it next time. Love, Linda."

In a subsequent class Peg said, "Do you know what happens to women's pelvic bones during a pregnancy? They soften. That's to facilitate passage of the baby's head. Then after the birth they harden up again."

And she said, "Colostrum. That's not been understood generally until recently. People thought it was just gunk that had to be sucked out of the breasts before the milk could come through. A sort of plug. But you know, our bodies don't do wasteful things like that. Colostrum. Full of protection for your baby, protection against disease. No medical lab could make it better." Roger never liked the sound of a baby breast-feeding, the eager squelchy sucking. Both of ours had gone on the bottle at three months.

And she said, "I'll teach you all I can about second stage. How to push. But really you already know. When the time comes, it'll seem so obvious you'll think, 'Why did she bother? Of course this is how I do it.'" Several women laughed, not in ridicule. I hoped with all my heart that this would be so for me, this last time.

And she said, "Pregnant women have really amazing endurance. Courage. Not only physical, but mental and emotional too. Women have coped with all sorts of awful situations for the sake of the unborn child, and borne it strong and healthy at the end too." She looked at her notes and went on to the next point. I did not move my own eyes, but could feel a couple of glances going past me to Grace.

And she said, "Everyone always wants to know, Will it hurt? The answer is Yes. There's no point saying otherwise. For some of you a lot and for some of you a little. But you'll know why. And you'll know what to do, or have done, about it." It was very good that she said this so matter-of-factly. I think that those of us who had already had children felt pardoned, in a way, and those who were pregnant for the first time felt relieved.

Roger wrote, "Tell me more about this course you're taking. Does the teacher treat you like a graduate student?"

I wrote, "Our teacher is English, I think from the north—Huddersfield? The exercises she is teaching us are based on the theory of psycho-prophylaxis. Some of them are quite hard to do. You'll have to learn quickly when you come home."

He wrote, "Are you really sure you want me to be there? I bet it'll just pop out, this time."

I could not visualize Roger in the class at all.

As I was leaving for class three or four weeks later, I realized that I had not changed my clothes. I was wearing jeans and the old shirt of Roger's that I'd had on all day. That night I saw that many of us were less dressed up, made up, than we had been at the beginning. I thought of dancers practising, the stripped-down look, hair bound back, total absorption in the movement of the body, in the discipline and understanding of the flesh. Also I liked the words, the terms, that we were learning, and found that several of the women shared my pleasure. We enjoyed using this gynaecological and obstetrical tongue with each other. Pride, really, it was. I tried to express this to Roger, giving him examples—dilation, effleurage, presentation. He wrote, "My goodness, this all sounds terribly technical," and I could see him sitting in the Tokyo hotel room, filling the crisp blue airletter before him on the desk.

I felt that Peg was preparing us, we were preparing, to give birth honourably, no matter how our labours turned out.

Yet one thing about the class distressed me, nibbled at the edges of autonomy.

The men. Why were they there?

Peg said, "You fellows now. You've got to understand that there's two of you having this baby. You're a team. It's all teamwork."

She also said, "You have to know all that I'm teaching as well as your wives do. Better, in fact. You've got to know the signs of transition, so you can tell her when to change to the special breathing if she doesn't do it herself. You have to notice if she's hyperventilating. Or if she's getting leg cramps. Or if she's becoming exhausted. The doctors will rely on you."

And she also said, "Remember, you're the coaches." She said this many times.

A lot of us found the C-level breathing hard to learn. It requires high control. As always, Grace had no difficulty learning it. She seemed to do everything perfectly, and she always had that disquieting passionate look, as if she saw something no one else could. Not something pleasant; occasionally her expression was close to horror, revulsion, strange looks on the pretty young face. She looked so young to me. She was very much alone. By now the rest of us women chatted in groups, before class and during the break. I tried, several of us tried, to include Grace, but she did not respond and we stopped. The only person to whom she related was Peg.

Garth—of course—also breathed C-level like a master on the first try. He watched the struggles of the class for a bit and then said, "Oh come on girls, you can do it. Can't they, fellow-coaches? Look, we'll show you," and he began loud C-level breathing, gesturing to the other men to copy him. Many did. Many of the wives looked at them anxiously, humbly, and tried again. I could feel exactly how I would have looked at Roger if he had been there. I looked to Peg for help but she was working intensively with one couple—deliberately?—and did not see what was happening. I could sense us women retreating into our separate selves, and I swear I could almost see bonds forming between Garth and the other men.

Then when we had to lie on our sides to practise breathing for transition, Garth gave Grace a heave over, saying, "OK fatty, here you go!" Grace did not react. But a lot of the men chuckled. Some of the women looked hurt, puzzled, and one had tears in her eyes as her husband exasperatedly "helped" her to turn over. I was glad that I had to make the move myself.

Then, in the sixth class, the childbirth film was shown. At the moment when the baby's head began to emerge from the mother's body there was a close-up. The silence in the classroom was complete. People held their breaths. Then Garth's voice broke in, saying chattily, "Bit of a butcher-shop look to the whole thing, isn't there?" And indeed there was a lot of blood about, but I am sure that most of us, women and men alike, had been seeing new life. It was a near thing. There was some relieved male laughter, but then Peg said, "Look, there's the shoulder turning," and it was, and there was a great rush and the child was born and the class fully caught up once more. Of course Garth's words still stayed in the mind.

Now here is the part that I don't know how to try to communicate to Roger. I can feel so exactly what his response will be: "But Linda, why didn't you do something about this man, if he was as nasty as you say?" Why indeed? Why didn't we? Particularly, why didn't I? I had age and experience on my side. I suppose I am a coward. I could imagine so easily how Garth would turn aside, turn around, anything I might say and transform it into emotionalism or prudery. And I did not know if the other women would support me, with their husbands there. And I could not think of what to say. It was hard, because Garth's manoeuvres were so quick. Seconds after he'd spoken and the class had reacted, he and his followers would be back in their serious student roles, breathing or counting or singing their D-level songs. A few times Peg said, "Garth, please. Attend to your own practising," or, "Garth, I think the class is doing just fine on their own." And he would say, "But Peg, I was only trying to help," and look aggrieved; and some of the men—and one or two of the women—would give little snickers, and then Peg's reaction would seem like a silly fuss. She stopped. I did not want that to happen to me.

I believe that most of the other women, and a few of the men, felt as I did, but none of us spoke up. It was something like this: These classes are about preparation for the birth of a loved child. That's *all*. They are not about confrontation, or rudeness, or mockery, or hatred. So even it there is someone introducing these things, we will not grant them validity by countering. I did not even speak to Peg, though I knew she would have welcomed my approach, welcomed an ally. That's perhaps what I feel worst about, that I didn't move towards her. But I did not want to be in that ruling role. And I suppose we all thought—I know I did—the usual things like, There are only a few classes left anyway, and We don't want to waste precious time with this stuff. But still what it amounts to is that we, I, did nothing. Not then. And therefore, I believe, other things had to happen. After a certain point they had to, as after a certain point an abortion cannot happen and a birth must.

Garth went on making his nasty interventions.

Grace looked at Peg with grateful love and otherwise behaved as though she were alone in the room.

We, the women, began to look angrily at Grace.

"She should do something."

"He's her husband, after all."

"She should speak to him." I myself said that.

"I think she's crazy."

"So do I."

And so we went further than stopping our attempts to include her. We avoided her.

Now we were at the seventh class. One of the couples in the group was Japanese, and neither wife nor husband spoke fluent English. Peg had got into the habit of spending the ten-minute break with them, going over the material just presented to make sure they had understood. At first the class smiled on her for doing this. Then I heard Garth in the smokers' circle say, "Seems a bit unfair to me. I mean why should the Japs, sorry, Mr. and Mrs. Nakamura, get all her attention? Suppose one of us wanted to ask her something now? They should have special classes for these people. Of course this is the first class Peg's handled all on her own"—how had he found that out? was it true, even?—"so perhaps she's just not able to deal with this sort of problem yet. Pity." That was the first time he had moved openly against Peg herself, and the excellence of her teaching was armour against his words; yet when people came back into the classroom there was a murmur all round the circle as the non-smokers heard what Garth had said. He looked smug. Mr. and Mrs. Nakamura had that awful look of people who know they are being talked about but don't know what is being said, except that it's negative.

Here again, you see, I think of Roger. How by this time in the class he would have known the Nakamuras—I didn't even know that was their name till I heard Garth say it, I had stayed at the friendly-to-foreigners smile level—in fact he would have known everybody's name. And instead of lying on my mat during the breaks, I would have been up with him, socializing in the anteroom, even through neither of us smokes. And how he would have said to Garth, robust and direct, "That won't do. Peg's meeting her responsibilities to *all* the people in the class, best as she can." And he would have gotten friendly with Peg and chatted with her about England; almost certainly he would have found a friend of a friend, or a place, or an occasion, that they had in common from their lives in Britain, and Peg would have warmed to him and felt some support. At least I think, I hope, that's what he would have done.

In the second half of the class that seventh evening the topic was baby needs—clothing, furniture, toys. Discussion showed that some new parents were buying as if there were no tomorrow—I remember doing that for Colin—and others were relying on borrowings and hand-me-downs. Garth's eyes glittered, and then somehow the issue became whether each set of parents was buying a new crib or using a second-hand one. It was dreadful, the ugliness, judging each other on the criterion of money. Boar-Neck's wife was almost in tears and so was I; I have rarely felt so helpless. Peg finally spoke, angry and nervous.

"Really, Garth, you make it sound as though people like us who don't have much money can't really love our children."

"But I said nothing of the kind!" he expostulated. "Did I?" and of course he himself had not. But on this matter even some of his most ardent followers were on the other side, and so he lost ground. Temporarily.

"Lot of silly extravagance, I'd say."

"Just for show."

"As long as the baby's warm and cosy, what does he care, eh?"

"Right dear," said Boar-Neck.

Peg then firmly moved the class back into session, and said we would take a few minutes for any questions, on any subject. Episiotomy was mentioned. Was this really necessary? Always? Evidently in the Scandinavian countries obstetrical staff were trained to ease the baby's head out slowly, gently, so that there was no tearing. Peg frowned, and then some of the women who had already borne children began to speak up. And I myself said, "Those damn stitches were so sore afterwards. That was almost the worst part, hardly able to sit down for days." And others were starting to tell similar stories, when Peg interrupted, speaking sharply as she had never done before.

"Well then? You want not to have an episiotomy, and risk tearing right from the vagina back to the anus, massive repairs, your body ruined, semi-incontinent perhaps? Well then go ahead. But if you take my advice you'll listen to your doctor and if he wants to do an episiotomy he should do it." She looked down at her lap and we saw her shaking. We were all quiet and afraid.

Garth said, "Trust your doctor, eh? Doctor knows best?" There was joy in his voice. "You'd agree with that, Peg?"

She took a deep breath. "I'm sorry I spoke as I did. But I still think what I said is right. Now please, just before we leave, let's have some breathing practice. For transition, please. Husbands, give your wives the count."

Garth smiled broadly as we women all obediently turned over, wanting to do our best for Peg—at least that's what I felt as I tapped the rhythm on my leg. We lay there, all us women, going, "One-two, one-two, hah!" And for a few moments everything began to feel right again, the learning was taking over. Then Garth began to parody our breathing, making sounds like the heavy gasps given just before orgasm. Boar-Neck snorted, a few men completely broke up in laughter, several women turned red, I could feel heat erupting on my own face. But again he stopped so fast and got his own people back into real practising so fast that some of the class didn't even grasp what was happening. The atmosphere was terrible, anger and discord and humiliation jittering and jagging as we got ready to go home. I could hardly wait to get out of the room. Peg did what she could, spoke to each couple in her best friendly-teacher manner, but I could see that although they were grateful many felt that the thing was simply too much for her. Damn Garth, what he'd said about this being her first class series. The cars pulling out of the parking lot were noisy with argument and tears. I was grateful to be home, so glad to be away from the shame of not having spoken. I drove home and reread Roger's most recent letter. "I don't care for the way most of the men here seem to treat their wives. They are extremely courteous but I think that is only on the surface. There doesn't seem to be the companionship you and I have."

All that next week I said to myself that I was not going to go to the next class and all the time I knew surely that I would. I thought about Grace. There must be some ruling reason why she had to stay with Garth till the baby was born. Money? She had none although obviously he did? Once I'd heard Garth say, "Since we've been in Vancouver," so perhaps she was friendless? (Who, after all, could I turn to if I were in the same boat?) I felt dreadful for the way I had behaved to her, and yet it did not seem as though it was wholly me, us, the other women, who made and kept up the barriers. Was it that she felt she had only strength to cope alone and could not handle any new connections, attentions? As if she had drawn a circle around herself and the baby. There was a quality in her which I found very moving. To say virginal of a pregnant woman appears stupid—but I remember it in myself and I have seen it in other women carrying a child for the first time. There is a purity of concentration, a fresh bloom. These meet their end in the birth and do not come again in subsequent pregnancies.

I wrote to Roger on the morning of the eighth class. "I'll be so glad to see you next Tuesday. My last class ends just an hour before your plane gets in, so I'll have good time to get out to the airport. Roger, when you are home we must talk about ourselves." Then I thought I should have written, "I want to talk to you about myself," but I did not want to cross out what I had written nor to recopy the whole letter, so I sealed and mailed it as it was.

The first part of the eighth class was low-key, after all my, our, agitation. The class was smaller, though; one or two couples were not there, and Peg did not make any happy announcements. We had visitors, two sets of parents from a previous class series and their extremely small babies. The sight of those two wrinkled snuffling infants was very satisfying, especially to the first-time parents. They were real, they had really come, after all the muscle relaxations and counting and breathing and exercising. Grace hardly blinked as she gazed at them. Peg seemed pleased at the happy attentive class before her, and then towards the end of the hour I saw her glance at her watch and run her tongue nervously over her big teeth.

When we gathered again after the break, Peg did not sit down with us in the circle. Standing at the edge, she cleared her throat.

"Tonight," she said, "we're having a dress rehearsal. We're going to start towards the end of first stage. Go through transition. Second stage. Pushing practice. Special breathing for crowning. Delivery. And to make it as real as possible I've brought some props." She went swiftly to a standing partition at the end of the room and emerged pushing a tall narrow stretcher with a thin green leathery mattress. At one end, metal stirrups clanked and wobbled. The delivery table was piled with masks, green gowns, draping cloths, gloves, packets of instruments. Everyone gasped, there was excited Oh look chatter. Peg smiled. Her lips were tight over her teeth.

"Now we need a mother. No girls, not you. I don't want one of you pushing too hard and suddenly doing the real thing. We need one of the fellows to play mother. I think we need Garth," and her long bony finger pointed straight at him.

There was a great Aaaah sound as we all took breath. And looked at Garth. He was completely taken aback and for a second did nothing and then it was too late, I was getting up, we were going at him, and we were all around him, he was surrounded by enormous bellies.

"A good plan." That was Mr. Nakamura.

"I like it, I like it!" That was Boar-Neck.

"Sorry, dear fellow-students," said Garth in his snottiest and most sarcastic tone, looking up at us, "but it's no go."

"O yes it is," said Peg, and we women moved in more closely around him and then he was up. We moved him en bloc, nudging him along to the table. He stumbled and I shoved him upright. I could hardly believe what I was doing. I turned to see Grace. She gave one long intent look at the procession on its way to the caseroom. Then she picked up a book.

Garth stood by the delivery table. His face was fish-belly white. The atmosphere of the room was terrific, silence pounding. I don't know if he got on the table by himself or if we pushed him, but there he was, flat on his back. We all got round the table and shared out gowns and masks and put them on. The gowns looked like giant bibs.

We got a gown and a pair of long green floppy socks on to Garth. I myself took his beautifully-polished shoes and threw them on the floor. Roger would have had a fit. One woman shoved a pillow under his head. One took his pulse. One listened to the fetal heartbeat. Peg said, "Contraction starting now," and we ran him right up through the breathing levels from A to D and down again.

"You didn't sing anything for D," I heard my voice saying sharply.

"Transition's just about over," said one of the women.

"Up then," said Peg, and two of us took Garth's feet and put them into the stirrups. He looked awkward, helpless, legs in the air and genital area wide open. Several women smiled, and I felt my lips opening.

"Contraction now," said one.

"He's not breathing properly," said another.

"They get confused in transition," said a third, "here, give him the rhythm." And we all chanted, "One-two, one-two, hah!" We stopped.

"Next time you can push, dear," said one of the women kindly.

The men watched the birth the way birds watch snakes. A couple of them were dragging on unlit cigarettes. The woman next me happened to glance their way in search of a syringe, and they flinched as if a searchlight had fallen on them.

Grace had laid down her book.

"Contraction now. Head back, deep breath, head down, *down*. And push. Push!"

"Down and forward and out. Down and forward and out."

"He's a rotten pusher."

"They all expect to have it done for them, these days."

"Shall we put him out?"

"No, make him push some more, he's just lazy."

"Come on now mother, we've got work to do, we can't do it all for you, you know."

"Here we go again. Deep breath, head down, *down*. And push. Push!"

"He's not getting anywhere with this."

"Epidural and forceps."

"OK."

"Now dear we're going to give you a little shot."

"Turn him over."

"No, wait till he's in the middle of a contraction, it hurts them more then."

"Over."

"Back down now."

"Now dear, you can't tell when a contraction's starting any more, 'cause you're frozen. *We'll* tell you when it's happening."

"Push now. Push." That was Peg.

Roger's face, Garth's face, I mean, very white now, a few tears on the cheeks even.

"No point snivelling, mama, you're almost there."

"What're you crying for? You can't feel anyway."

"There's the head."

"No more pushing. Gently next time, be crowning now."

"He doesn't know what that means. Listen dear, just don't push so hard now, OK?"

"Shallow breathing, lightly lightly, that's right."

"Soon your beautiful baby will be born."

The whole class was now stiff with excitement waiting for ? to be born. Birth, birth, new life, the excitement, the fear, will it be all right? Oh please let it be all right, let it come, be born now, oh be born now, now, new life be born.

"Head's all out. Gently now."

"Next time'll be it."

"Shoulder's turning."

"There!"

We all moved back from the delivery table a bit, and the woman who had been bending down close to Garth's crotch stood up, triumphant, cradling in her arms—nothing, of course. The emotional temperature in the room fell about thirty degrees. She turned uncertainly to Grace.

"Did you want a boy or a girl?"

Our fantasy disintegrated before Grace's expression of nauseated loathing.

The doctor uncradled her arms and took off her medical accoutrements, and the rest of us did the same. Garth got his feet awkwardly out of the stirrups. His hair was messed, his Pierre Cardin shirt rumpled. Silence. Hatred. Rage. Some of the men went to get coats. Garth walked over to Grace and sat down by her to tie his shoelaces. She looked at him.

What a look.

With gravid dignity she rose and walked out of the room. The child went before her. Peg took a step after her and then stood still, concerned.

We women were still red-faced and panting.

Garth got his coat and buttoned it up. There were twin threads of saliva at the corners of his mouth. He went shakily out.

"Does he ever look crazy," said Boar-Neck. People blinked.

Today I got a letter from Roger. It seems he unexpectedly spent the last two days of his trip out of Tokyo, so he won't have got my last letter.

Tonight was the last class. Grace and Garth weren't there. Peg didn't say why. The group was low-key and attentive. People smiled at each other gently, carefully, and some of the couples held hands. When the lesson was over, Boar-Neck thanked Peg on behalf of the class, and we all clapped. Peg blushed.

Then came the surprise. The men had arranged for flowers to be delivered to the classroom, and so there were red roses for each mother-to-be. Some of the women cried.

The road to the airport is busy as usual. I have to think hard about my driving, for I'm awkward behind the wheel now that I'm so big. But I can feel the roses crimson on the seat beside me, and soon Roger will see them, and I shall have to try to explain.

The Animals In Their Elements

There came a time when Harry's parents "couldn't manage" their big house any longer. After anxious consultation, they installed plumbing in an awkward closet off the front hall; they put the double bed in the dining-room. They never went upstairs in their own house again. Never. Their daughter-in-law Shirley had been contemptuous. "Rats in a trap," she had said to her husband George. Thirty years later, though, she cried in terror, "And I've come to that now, in my own house! I won't, I won't live on one floor like a rat!" George's brother Harry was not sure that rats lived so; immediately he saw in mind his arcing birds; but George got out the Yellow Pages. Workmen came. Soon, chairborne, Shirley rose in electrified stateliness from the first floor of her home to the second, and then descended. The strong chair hummed thus for the twelve years till Shirley died. Then, after a while, the old brothers took to piling things on it that needed to go upstairs or down, although neither George nor Harry actually rode in the machine. Princess the cat liked snoozing there. Perhaps the leather seat smelled good? But if a hand reached for the switch, she jumped off. Following George's death—he got paler, less there, after his wife died—Harry did not use the chair at all. The deep resonant thrumming felt too loud in the quiet of the old house.

In time, in time a brief rest at the half-landing wasn't enough when Harry went upstairs. His heartbeat's volume made him shrink. Would the creature beat right through the chest-wall, pulse out into his hands? He stood and waited for quiet, sometimes hearing the mice as they skittered and squeaked companionably under one particular tread. Harry smiled. The veins at his wrists resembled the guywires of the phone pole in the back yard (the birds wound their claws round the strands so neatly), but surely his thin arms did not need so much support? Perhaps he was getting hollow-boned like the birds, for the bathroom scales George had bought forty years before had never given trouble and they said that there was, each Monday, a little less of him. Bird-bones. Thin as grass they must be. The radio once said that the Indians used them as—flutes? Birds ate a lot. Harry ate as always. The green beans were frozen now, not canned, the cottage-roll shrink-wrapped in plastic, but cornflakes and baloney, fish sticks and ketchup, boiled eggs and canned peaches were the same. Sometimes when Harry rose from the table a blurry dizziness vibrated in his head, also sometimes when he was on the toilet. Number two was all right. It came only while he was upstairs near the bathroom. Number one became more and more insistent during the day. When he finally could not struggle up the stairs more than once daily, the kitchen sink presented itself. Mother, Shirley would have been appalled. Princess patted curiously at the yellow stream. Harry scrubbed the sink hard after.

Cat and birds. Each spring at least one nestling died the death of the teeth, no matter what hours Harry guarded the maple and the old walnut and the swallows' place under the eaves, no matter that Princess's old legs were stiff and her eyes smeary. When Harry cast seeds, crumbs on the grass, Princess meowed so loudly behind the dining-room window that the birds would not come; with the door open a slit as Harry came back in, she leapt out into the ocean of tall tangled grass, hissing to get at her prey. He held her on his lap in the evenings, while the radio talked and sang to itself behind his shoulder, and told her over and over she didn't need to do that. Didn't he give her Puss'N'Boots and milk and chow every day? didn't she even leave food uneaten sometimes, a terrible waste? His fingers smoothed over and over her aches; she purred raspingly, nosed his chill hand, put her paw on his veiny wrist. Next day, the same. (But the crumbs and seeds were gone. Sometimes slugs came—so fast, after rain, that perhaps they grew up through the wet dirt? Yet when he watched them journeying they hardly moved.) And Harry told Princess how beautiful the birds were, saying the word again and again, though it didn't show how they darted, swooped, turned, spiralled down. And they did their business in the air, on the wing, a little squirt and a splat on the window or porch rail, all done. Most of all Harry could not find how to tell Princess about when a bird landed nearby and looked at him. When that happened the rest of the day was different. Oh, their eyes.

The pulsing feathers on the throat, the legs sheathed in tiny fish scales—but mostly the eyes.

Harry also worried about how Shirley and George had done so much work around the house. After he had swept hall and kitchen, wiped counters, run the carpet sweeper, and used Dutch cleanser on the plumbing he could not think of anything more, but he felt the house did not look as it had. Perhaps, though, Shirley had cleaned the rooms that were not used? That was like how she and George used to talk about what they called old times. Jobs, visits, parents, occasions. Harry did not see why. His job had only been the place he was in when he was away from home, first Mother's and then Shirley and George's. Years in the basement stockroom, cartons and files; forty-seven years; she would have been proud, not a day sick. And George had always fixed, painted, contrived, right to the end.

Shirley had also made lists for George to take to the grocery, the laundry, but Harry knew the Sunshine Market's layout by heart, and picked up his supplies weekly in the same orderly route round the store. Sometimes there was a confusing new product on the shelf. Then he went to the back and found old Mr. Gilbert, who gave sharp instructions to a grandson or granddaughter. Mrs. Chang also had grandchildren—possibly great-grandchildren, with these people it was hard to tell. As polite as she, they gave and received his laundry parcels with broad smiles he liked. Once he felt faint there. They sat him on a spindly chair and gave him strange tea. Mrs. Chang laughed and talked a great deal in Chinese when he took a second cup. He needed no list there either.

Time's altering puzzled Harry most. Waking just before the old Westclox went off at seven-thirty, he enjoyed Princess's warmth by his feet, or watched her crouching at the baseboards listening to the life within the walls; she lashed her tail and clicked her teeth, t-t-t-t. Harry lay in his high bed looking out over the back yard, and thought of the nests of birds. He had seldom entered Shirley and George's room, never since his brother's death, and there was no reason to go into the "guest-room" where Shirley had done her mending, none to struggle up the final staircase to the narrow black attic. Once he had gone downcellar with a man from Hydro to check the furnace. Beautiful moths there, giant spiders with freckled markings like maple walnut ice-cream. There was a path worn on the hall carpeting. Did birds have roads, in the air? Now somehow it was past eight. Bath. He feared that he might not be able to get out. Not daily might be all right? or sponging on the mat before the sink? But since childhood there was bath, not wastefully deep of course, but under, and sometimes the thought was almost felt in Harry that certain parts of his body welcomed that immersion. Then dressing, downstairs, let Princess out ("She's coming, birds!" from the back porch), cornflakes and toast. The last of Shirley's home-made jams was sadly gone. He missed her baking too. Once Harry had opened the cupboard where her flours and grains and powders were stored, and there came a great soft fluttering of minute papery wings, beige and brown; he closed the door. Tea steeped while he ate, strong enough to trot a mouse as Mother always said. Ten-thirty. How? Clean up after himself, always. His tasks done at last, he could watch the birds, front porch or windowseat depending on weather—but it seemed he had scarcely sat down before the noon O *Canada!* went off downtown there. Sometimes Harry was breathless with speed.

So he no longer attempted two errands in one afternoon. Friday became Bank, and Tuesday Laundry. Sometimes there was Drugstore. His parents hadn't had much luck with doctors, Shirley had despised people who went to them, but Absorbine Jr. made Harry's arms feel less stiff temporarily and the Neo-Citran the radio said about was nice before bed. Often though there were unknown sales clerks, like the changing tellers at the bank. And like the mailmen. The old fellow had stopped a while after George died, and they couldn't seem to settle on someone new. Right now there was even a girl. She had curly red hair and wore shorts sometimes. Every morning she noisily thrust his meagre items through the slot and jumped back down the steps two at a time, and by then Harry'd made it to the front door, and every morning they waved and smiled at each other. He felt he would miss her if she were to be reassigned, even though he'd never seen her up close. She seemed a long way away, down there on the sidewalk. Commercial Drive, where all his errands were, also seemed farther away than before. A great many people now walked very fast there.

One day at the bank the vibration visited his head. He feared he might have to leave and try again next day, or ask for help—there wasn't a spindly chair. Neither of these dreadfulnesses happened. He did not even ask to move ahead in the line. Pride got him home. Next day he slept through the Westclox. Almost nine—worried, ashamed, Princess figure-eighting and meowing about his legs—he got himself unbathed downstairs to let her out. Then the hallway stretched before him like a bone's black tube, dark and curving-down and terrible, and he was sucked head-first into it while birds squawked and hit the air about him.

He did not feel like talking. Nothing new, although he'd liked to hear George and Shirley's voices going on together. Homelike. Here was not home. These unknowns were nurses and doctors, that was all. There was nothing to be said to them. He must just work hard, very hard, to get better. One young doctor, Jewish he must be from the nose, had dark bright eyes, and Harry liked it when he visited. There was a sandbox in which he must feel for letters, to match them with letter-shaped holes in a container. The sand felt silky but it got under his nails. He worked hard, got pretty good at the matching. Now an exercise book. C dash T, put a letter in the middle to make a word. Impossible. No such. Then, as when a bird lands on a rainy branch and the drops frill off the leaves on to those below, there was an infinitely small touch in his head and the letter A presented itself. He smiled. The young doctor smiled too. Then Harry wept, and pointed to the word and wept. Some later time a nurse came saying, "He says to tell you Princess is all right." Belief and incredulity nested together in his mind. No speech. Birds flew distantly over the parking lot beyond his plate-glass window.

Some of the food was surprisingly familiar. Shirley had made jelly like this, with bits of fruit suspended in it. Harry wished he knew how, and how to make his left leg work. He thought he was doing his best, but nothing happened. At first young girls helped him to the bathroom but of course he could not go, and finally they brought a man. People got him to move his limbs in a way he saw, but did not yet feel, was rhythmic. In time, in time the leg felt more, though it was ponderous, dragging, with dry skin shaling off like fish scales under an invisible knife. Once, perhaps a dream? Mrs. Chang was there, grinning at the end of his bed with a paper bag. Later, fat pale buns on his bedtray.

Mr. Gilbert strangely drove him home, and on the back seat Princess was angry in a cat-carrier he had never seen before; her paw stuck out of a hole. At home she washed and washed, then slept so deeply he was fearful. He himself could not get up the stairs. He tried for a long time. Then the doorbell rang and there was the woman. Eventually, after tears, Harry saw there was no other way. He must now use the chair. Princess got on to his lap and did not move even when he pressed the switch, and he cried and loved her. The woman walked beside him up the stairs, distracting, but he got to the bathroom all right (number two at midday, he was all dis-ordered) and then back down again, still with her beside him, talking, and she came into the kitchen with him, talking, and she opened the refrigerator and frowned.

This woman came in the mornings. She helped him dress and prepare breakfast. She worked about the house, did errands, got lunch and left. Then for two days running she didn't come at all. Relieved and upset, he found things difficult. Then she came again. As he watched birds clustering on the phone lines, so still and yet a millisecond from full flight, "weekend" arrived suddenly. When she arrived next he smiled at her, and she talked and laughed and gestured. Harry liked that. Since he did not get out any more to see Mr. Gilbert or Mrs. Chang, or their grandchildren, or the unknowns at the bank and drugstore, it was nice to have this woman to look at, to hear. The first time he saw the mailgirl that was nice too, for she ran right up the steps and talked and smiled with the woman there. The smooth skin on her legs shone when she ran down again. When Harry walked there was no rhythm, just slow thud thud; but once he got to the back porch he could stay there longer now. Crowds of birds all suddenly flew into a tree, so it was full of movement and sounds and leaves, and then they all flew away together like blown blossoms. The lilac had bird's-eye beads of rain shining all over the purple fragrant cylinders. The grass in the yard, uncut for years, was rich in small living things, and the land-birds dove into it like cormorants into green sea. When it got dark he must go in, check the oven and fridge, because she left supper-foods there and if they were untouched her face got wrinkly and her voice hurt his ears next morning. He and Princess hummed upstairs together in the dark; he had trouble manipulating light switches, and for the same reason no longer used the Westclox, though he feared sleeping too late.

Several times the woman got him to the front door and pointed down to a blue car. Grasping the doorframe, he resisted. Her voice got loud, he wouldn't, he didn't care if tears came. Mr. Gilbert and the mailgirl arrived, talked; Harry looked away, where branches swayed under the landing weight of a bird. Then, changes. She began to make him exercise each day, as in the hospital only more vigorously now, and although Mr. Gilbert generally gave her the right things to eat there were new foods too. There was a brown bread with hard bits in it, a cheese much stronger than Velveeta, a plastic cup filled with a sour junket. At night he put these in the garbage. A terrible waste. Fearing discovery, he buried them as deep as he could.

Once after so doing he went still angry up to bed, and realized that he did not have to stay upstairs at night. The chair would take him down and bring him up again, and no one need know. He took a blanket down, and went laboriously out into summer night on the porch. She had trained sweet peas up the old trellis there, and there were no more sharp points on the old wicker chairs. He settled in one and breathed the sweetness. A spider's web veiled the moon. Princess, enraptured, vanished into the dark. Harry called her for a while and then sleep took him. Some shivering time near dawn he woke, achingly rigid and with a swollen bladder. Using his left leg felt like lifting a bag of cement. He urinated over the railing and watched the pool sink gradually into the earth. There was a slug's silver trail on the pathway, heading for the house. He heard the very first curoo of the day from a pigeon under the eaves, and saw Princess come pad-padding through the dewed grass. He was days recovering, the woman's face and voice did that, and he knew he could never repeat the act.

Harry looked about his house, and saw that its aspect was different. He had been right then; this woman, like Shirley, knew what to do. Windows shone. The meal-moths were gone from the kitchen cupboard; he regretted that, but he knew it was proper. He could see the pattern on the hall carpet quite clearly. The cushions on the living-room chairs were all fat again; although it was not his custom to sit in there, he liked to see them thus. Magazines lay on the coffee table, too. He did not look at them, but he remembered Shirley and George doing so, and when sometimes the woman did while she drank mid-morning coffee, he liked to see that. Then there came a magazine with a front-cover picture of happy birds pecking at a big ball that hung from a ribbon. The paper was shiny so the birds' eyes gleamed. He showed the woman. She riffled through the pages and found a series of drawings.

"It's how to make a feeding ball," she said, "suet and peanut butter and seeds. Birds love them."

"Can we make it for them?" he asked. She dropped the magazine.

The old yellow bowl he remembered his mother using gave off a rich, oily, choking smell as the woman measured and ladled and poured and mixed. He watched the stuff coagulate, and suddenly she said, "You too, Mr. Eldridge," and she put his startled fingers into the slithery dripping sticky grit-pocked sludge with hers. He could not bear it, but as he moved his hand the mass gave, pleasingly, beneath his pressure. He began to squeeze and squeeze, and gradually between them the thing took its form; still he could scarcely wait to wash and wash his hands. She hung the seed ball where Princess's greatest imaginable leap could not reach. Then came days, days, days of birds. Harry's eyes were filled with gloss and floury softness of feathers, brilliant glances and quick nervy turns of the head, speckle and stripe and the way one iridescence in a moment became another. He watched, he loved, for hours.

Deeply grateful, Harry felt his heart become swollen gradually with some disturbing awareness. Then it came to him that he must do something for the woman in return. He knew, immersed at once in fear, what she wanted. He would have to leave his house and his birds and Princess and get into that blue car.

Once, years before, when Shirley was rising on her chair, the folds of her skirt had lain so that unwitting he saw past her knee. Movement came between his legs. If only he could be on her as she rose, in her, fitting like a fish in dark water to nose and wriggle—and then were hours and days of hiding the shame, in his room, or thankfully at the work where he did not have to see Shirley or George and could hide that bad thought. Now, as then, Harry skulked, fighting his knowledge, sullen, eating less, resisting the exercises. Out his window, birds dropped and soared; with screams and squawks, a group encircled one of their number and attacked with a loud throbbing beat-beat-beat of wings. Harry turned, almost fell, wrenched himself down the stairs on foot. He got to the woman at the front door and pointed out to the car, nodding.

There was to be a special morning. Yes, there was to be a special morning, but he was not clear when, and so woke several times in agitation. Then, "These are for you," she said on arriving, and when he got the package open there were dark blue undershorts, only of thicker cloth and with a string to tie. Without stopping she took him away then, down the tall steps to the distant sidewalk and inside the car whose doors were open, waiting to close on him. Not a long drive. A low building. Old people in wheel chairs, younger people pushing. Harry could walk. Inside, a strong smell that scraped his lungs, a big room with benches and padlocked metal cabinets, and beyond an arch lay a great rectangle of rocking turquoise that shone with bursts of light like a peacock's tail.

He was naked, and then had the blue shorts on over his dry scaly legs. The tiles were nubbly under his cold feet. Then came a ramp into the water, warm, surprisingly like a bath, but deepness was coming at him, not bath, and she would not let up, she was making him go further in, and further. He glanced down once, to see his body shimmer and angle away from where he thought it was, and the water and air about him were vast trembling territories. She made him grasp a rail. Suddenly the translucent blue took him up and his legs were somewhere out behind, turning, sliding, and everything under his bathing suit floated free.

Draw right back now and see old dried bone Harry, his skinny shrunken limbs fluttering down there in the corner of the tilting blue with the lights and the Muzak pouring over. His home care worker is by him, a comfortable middle-aged woman in a modestly-cut bathing suit. He turns to look into her eyes and says, "I am flying. What is your name?"

Neighbours

When Mrs. Grainger told them to, the boys sat down. The pale green paint of her kitchen chairs felt cool on their thighs. All the kitchen was the same pale green. The stove had spindly legs, the table was bare; no kitten could hide beneath either. The tall high cupboards were shut. No dishes, no knitting or bowl of fruit or open book lay on any surface. In the middle of one wall was a door about which Jerry was deeply curious. Short, narrow, closed with a hook, this door was so high he would have to climb on a chair to reach it. Surely no grownup could go through that.

With their bare plump legs hanging above Mrs. Grainger's kitchen floor, the brothers, four and five years old, gazed down at their canvas summer running shoes as though these articles held truth.

Mrs. Grainger made them tart strong lemonade by taking a Mason jar out of the fridge and pouring translucent syrup into glasses. To this she added cold tap water. The boys loved to watch the water spatter down into the deep white sink. Mike liked Mrs. Grainger's lemonade best. Jerry preferred Miss Jardine's, perhaps because of the jug. Made of blue glass, with high round shoulders, it stood ready in Miss Jardine's fridge, and after sitting out for a while, it grew a necklace of shining beads. Both boys preferred Miss Jardine's oatmeal cookies. Unlike Mrs. Grainger's, which were uniform, Miss Jardine's were all different sizes, each with its own charms. The big cookies had crisp edges and were sweetly jagged in the mouth, while the little ones were just wads of raisins lightly mantled with the soft oat dough. But sometimes at Miss Jardine's there was no lemonade, no cookies, only a glass of water the same as they could get at home: "Busy today, gentlemen, bye bye!" Sometimes when they knocked hopefully on her back door in the heat of an August noon she was not even there. Mrs. Grainger was always there, and always had cookies in the grey stone jar by the window where the green ivy dangled and glinted in the sun.

The two houses also differed. Jerry had thought that Mrs. Grainger lived alone in hers; it was so quiet, so the same always. If the three green china elephants that stood in graduated sizes on the kitchen windowsill were his, he would change them about every day, put them more closely trunk to tail or have them rush tusking at each other. Mrs. Grainger never moved them. Mike knew better than Jerry. He had seen Mr. Grainger coming home from work, seen him walk up the front steps and use a key to open the big door. "He had a hat bigger than Daddy's. And a big cane. And a cigar." Then, just recently, when the boys had been out with their mother, she had stopped as she often intolerably did to talk to people she knew, and suddenly asked sharply, "Well boys, can't you say hello to Mrs. Grainger? Think how nice she's been to you! And here's Betty, too." Mrs. Grainger looked quite different with no apron. She had put red on her thin lips. The other woman, grown-up but younger, gave the boys an uninterested look and went back to her shopping list. Jerry grasped that Mrs. Grainger was this grownup's mother. So Mrs. Grainger and Mr. Grainger and Betty all lived in that house. Still it felt lone, like the lone lad in the Irish fairy tale, though there were books and pictures, and coatracks, and the lemon syrup was by no means the only thing in the refrigerator.

Miss Jardine was by herself, yet her house was incomparably livelier. First Mike and Jerry had their lemonade and cookies, and told jokes. Miss Jardine laughed. Then they went into the living-room, where each boy could choose. The options were the shelf in the glass-fronted cabinet, the chess set, and the Russian doll. There were never fights. Jerry loved the Russian doll, who came apart to reveal a smaller doll who came apart to reveal a yet smaller, and so on down to a tiny scarlet and gold nub that didn't look remotely like a baby. Mike usually went straight to the shelf, on which stood, among a crowd of demitasse cups, a Chinese ginger jar full of tiny silver spoons, their handles carved and chased into dragons, griffins, swords, helmets, palm trees, beautifully-feathered birds. Occasionally for politeness the boys jointly chose the chess set, and set up the gleaming warriors in arrays dictated by the design of the Oriental rug. Later, each could shake the snowstorm with the little dancer, blow the silver whistle, look into the kaleidoscope, and count the eyes on the peacock fan.

Mike was always ready to go before Jerry. The younger boy was sure that Miss Jardine could show them more wonderful things if they stayed, but Mike wanted to go back to whatever game they had been playing before the desire for lemonade came upon them. No such conflict occurred at Mrs. Grainger's. When they had finished their refreshments, they went. Mrs. Grainger invariably said, "Now boys, you must go straight home and brush your teeth after all that sugar. Will you promise me?" They never once kept their promise. Mike suffered over this more than Jerry. Miss Jardine showed no interest in their teeth, often saying good-bye with simply a friendly wave, but sometimes she hugged and kissed them. They they had to feel the bones of her thin arms under the slippery dress-sleeves (not at all like the crisp cotton their mother wore), had to smell the powder on her cheeks and see the dark flakes on her eyelids. Jerry told his mother that Miss Jardine looked like a deer. "I can't understand the child," she said to the children's father. "Miss Jardine's blonde, with a round face." "Maybe he means 'dear,'" said their father. Evidently neither of them saw how Miss Jardine's full lips were framed by little gold hairs. Jerry especially did not like the holes in Miss Jardine's ears. When she wore earrings that was all right—she had a pair like new moons with stars caught in them that he longed to touch—but when her earlobes were bare the little slits looked terrible. He imagined the point of his mother's paring knife going in.

Mike and Jerry's house was between the Graingers' and Miss Jardine's. Only some perfunctory hedging and fencing separated the back gardens, and the boys roamed all three that summer of 1949, making the most noise in their own and devising quieter dramas for their neighbours'. Miss Jardine's had a well-designed plum tree, into which they hauled up toys and supplies to furnish a sitting-place. The Graingers', though treeless, was lined with flowering shrubs. Within these mature plantings were dark spacious rooms, coolly hidden from the sun and replete with insect life, where Mike and Jerry stayed for hours. If they went in to their house for home-made popsicles their mother, hugging them, would sniff their t-shirts and exclaim, "Have you been underground?"

When in the bushes, Mike was anxious to be invisible. He dragged fallen branches and leafage about chosen shrubs so that even a peering enemy would pass by deceived. This obsession suited Jerry well, for when not engaged in beetle analysis he liked to lie flat and look up through branches to the sky. Lying thus, Jerry heard many noises. The beetles made small tocking sounds as they fell from one leaf or dry twig to another, and the leaves themselves slid past and stroked each other continually. The birds talked all the time. Once Jerry heard a very small thudding sound, repeated, repeated. He lay still with his eyes shut, trying to feel what it could be, and when at last he slowly turned his head he saw the nose of a mole pointing up from a silky black cone.

There were human sounds too, of course: radios playing, bits of grown-up conversation from the lane or the driveway, the little girl three doors down crying again—she was learning to rollerskate—and the tinkly note of the ice-cream man. One day came a new sound. Jerry thought of ice-cream but immediately knew he was mistaken. He thrust his head out through the shrub.

"Get back in!" Mike shouted.

"Listen, Mike!"

This music was different from any that Jerry had heard before, unlike the radio, unlike the record of children's music that he and Mike had, unlike the bells on the church at the corner that his family didn't go to but the Graingers and Miss Jardine did, unlike the amazing toilet paper roll at the cottage they'd been to earlier in the summer—Jerry smiled at the memory. This new music stopped, right in the middle of a bit; a man's voice said something and the music started again, almost at the same place but sounding a bit different.

Someone was making this music right here and now.

"Jerry, get back in. You're spoiling it." Mike's face was red.

Jerry came right out of the bush and headed for the front of the Graingers' house. Mike opened his mouth to yell and then closed it. Jerry did not often have that look.

On the Graingers' dark green porch, latticed and shaded, sat two people, neither of whom noticed the brothers sliding past into a large forsythia by the steps. One was the Betty woman. She did not look as she had in the grocery store, but smiled and held a cigarette, red and smoky at the tip. The other was a man who held on his lap a thing from which the beautiful sound came. Jerry knew what pianos and violins looked like; this was neither, but a box with a curve on one side and strings across it. The man's hands slid over these strings so quickly that his fingers blurred before Jerry's eyes. He smiled as he played, smiled at Betty. Behind the green leaves Jerry smiled too and began to clap softly in time, but Mike snatched at his fingers. "Sssh!" Mike was right. So the little boys silently listened. The man played music that was sadly happy or happily sad, and quicker music that felt like marching with flags and drums, and music that made Jerry think of the giant cookie that rolled and rolled and rolled away over the hills and dales that lolloped across the broad pages of the storybook. Then the man stopped. His hands lay still, and he and the woman laughed. Then he lit a cigarette too. Jerry stared at his own hands.

"That instrument is an autoharp," said Miss Jardine next day. Jerry, halfway through the Russian doll, put her down because Miss Jardine's voice was strange. "Yes, it is very pretty music." Mike and Jerry looked at each other, confirming: Miss Jardine did not like the autoharp. Grownups lied often, but Jerry could not imagine why this lie.

At supper the boys sat with their parents on their own back porch, eating tender buttery corn and listening to the autoharp.

"Boyfriend here again, eh?" said their father.

"So it seems," their mother answered. "Jerry, you don't need any more butter."

"Wonder what'll happen."

"I can't imagine," in the tone which meant both disapproval and a refusal to discuss further in front of the children. "Don't you want another piece of corn, Mike?"

He did not. Further questioning revealed that he felt unwell. Jerry saw alarm on the adults' faces. The bustle ended with Mike tucked up in bed in his parents' room receiving a visit from the doctor, for August was polio season and every parent imagined the shrivelled limbs or small coffin that a fever might foreshadow.

Although Mike only had summer flu, for some days household patterns were disrupted. Waking one night after a couple of hours of sleep, Jerry heard voices in the garden below the bedroom window: his parents' voices and Miss Jardine's and that of an unknown man. He got out of bed and started downstairs, and on his way had what he knew was a good idea. He wished he could tell Mike, but went on alone, out the front door and round the side of the house to the point where the driveway diverged, towards the Graingers' garage on the right and Jerry's on the left. Here there was a large philadelphus. The little boy hid inside it, sniffing the warm night for a trace of the rich orange fragrance this bush had had earlier in the summer when it was covered with white flowers. None remained. The earth smelled sharper than in the daytime. Then came a whiff of something else, something attractive yet nasty. After a moment Jerry remembered its name. Cigar.

Miss Jardine, about to leave, leaned on the fence, smiling through the dim light at Jerry's parents and at the cigar-smoking man. He was rising from a lawn chair; his back was to Jerry and he looked extremely tall.

"Good night, Miss Jardine," he said. His voice was deep and firm, yet Jerry thought of the autoharp.

"Good night, Mr. Grainger. Good night, Alice, John." Jerry always found it strange to hear his parents addressed thus, stranger still when his grandparents visited and his mother said "Mum" . . . but this was Mr. Grainger, the unknown. Jerry wiggled towards the outer branches of the philadelphus. No, Mr. Grainger was not wearing his hat, but the thick cane of which Mike had spoken rested on the arm of the lawn chair.

"For a week. She'll be back on Sunday," said the deep voice, in answer to a question from his mother.

"So you and Betty are batching it." What did that mean?

"Yes indeed. Very pleasant to be just father and daughter for a while."

"And Betty's young man is about a good deal, I see." The little boy knew by the way his father's head turned that his mother should not have said that. Why?

"Yes," said Mr. Grainger, and his cigar went fiery red. "It won't quite do, of course. But I think it'll work itself out, if she's a sensible young person and if nobody gets too excited about it." Why did somebody get excited?

Jerry's mother said suddenly, "He does play beautifully, I must say. I love to hear him."

Mr. Grainger and Jerry's father laughed, the way grownups sometimes laughed at Jerry or Mike, and their mother protested. Then all three adults laughed and their voices changed and they began to discuss the thing called the paper. Someone had always seen something in it. Jerry's feet were getting cold. What was a Marshall plan? His father could go on for a long time about this. Jerry withdrew.

Mike was not awake. Disappointed, Jerry went back to their room and peered out the window. His mother was saying, "Yes John, let's, and I'll get some cheese and crackers out." She and his father vanished into the house beneath him. Jerry watched Mr. Grainger unfold himself out of the lawn chair, stretch, toss his cigar into a flower bed—his mother would find it next day and be cross—and walk to the side of the garden, where he stood looking towards Miss Jardine's house. Then Jerry heard his mother's checking footstep on the stairs and rolled into bed.

Because Mike's spoons might be lonely, Jerry played with them next day, laying them out on Miss Jardine's pink plush sofa that he could not make up his mind about. Was it bristly or smooth? Was that a nice feeling that ran up his spine, or not? The spoons shone, silver against the pink. Did they like the plush? Gently, Jerry rubbed his cheek on it, smelling cigar and watching the silver glisten. Cheek felt definitely better than fingers. How about mouth? But Miss Jardine came and sat down near him.

"Why don't you like the autoharp?" asked Jerry, looking up sideways with his cheek still on the plush.

"I do, Jerry. I told you, it makes a lovely sound." Miss Jardine's fingers moved quickly in and out of the lacy circles she was making. Like large permanent snowflakes, these rested on all her living-room furniture. Jerry thought of Mrs. Grainger's bare wooden chairs.

"Does Mr. Grainger like the autoharp man?"

Miss Jardine's fingers moved even faster. Then they stopped.

"Jerry, this will be hard for a little boy to understand. He is not right for Betty. If he played a violin, or a cello perhaps, in a real orchestra . . . but he just plays with some friends, in a band, for dances. He doesn't have a real job. Betty is the Graingers' only daughter, and of course they want something better than that for her."

"I like him," said Jerry. He took his cheek off the plush and gathered up the spoons.

Mike was better. They sat inside the Graingers' forsythia bush on Sunday and watched Mr. Grainger come out of his house. Although they could not see Betty, he was talking to her.

"And then I'll meet your mother at the station."

"When's her train due in, Dad?"

"Two, if it's on time. You should be dressed, Betty, at this hour, even if you are not attending service."

Mr. Grainger went firmly down his steps. He had his hat on and carried his cane. He struck the sidewalk with it at every other step. He went to Miss Jardine's house and rang the bell and when she came out he took off the hat and bowed. Laughing, Jerry and Mike bowed at one another all over their front lawn for some time after Mr. Grainger and Miss Jardine had disappeared round the corner and the church bells had stopped ringing. Then the autoharp man came, but he did not have his autoharp with him. He leaped up the steps of the Graingers' house and in the open door. Jerry and Mike saw Betty holding it for him. Then their mother called them for lunch.

Afterwards, they started trying to leap up their own front steps. Mike could jump up two steps, but lost his balance if he tried to do another two right away. Jerry could not even do that. In trying, he fell and cut his head. When he had cried and his mother had washed the cut and put Mercurochrome and a bandaid on it—which he fought, knowing how the bandaid would feel coming off, bringing hair with it—he and Mike went outside again, with popsicles, and sat to eat them on the front lawn. A taxi came down the street. The boys watched as Mr. and Mrs. Grainger got out. The taxi-man produced a suitcase.

"Hello, boys."

"Hello, Mrs. Grainger."

"Be sure to take your popsicle sticks inside when you're through. Your mother wouldn't like to find them lying on the lawn, would she?"

"We will, Mrs. Grainger."

Mr. Grainger already had the suitcase up on the porch and was pointing his key at the big front door.

Almost as soon as Mr. and Mrs. Grainger disappeared into the house, the shouting began. The deep voices and the higher ones criss-crossed. Unable to make out words, the brothers got up and went over to the Graingers' front walk. The noise was nearing the front door. It opened and the autoharp man staggered out. He did not have his shirt on, not even an undershirt. He was holding his shirt in one hand and with the other he pressed his cheek, on which there was a scarlet welt. His hair was messy. Sweat was bright all over his face. He came thundering down the steps so fast that the boys thought he would fall, but he did not quite and passed them as though they were not there, gasping in a painful high-pitched way. Looking after him, the boys saw that there were red stripes across his bare back. Crying sounds came briefly from inside the house. Then the door closed with a huge slam, the kind they were absolutely never allowed to do.

There was corn again for supper, and Jerry did not feel like having any more.

"Double damn," said his father, "here we go again."

"At least you're out of the house all day. I'm the one who has to cope with it."

His father gave his mother a big hug. Jerry always liked to see that and he could tell that she liked it too, even though she pretended to push his father away. Still Jerry's forehead hurt, and he felt trembly.

"Come on then, big boy," and his father picked him up. "Up to our room with you. I'll put the cot up and we'll have the doctor in again just to be sure. No, I'll phone him, Alice, you finish your supper."

Jerry woke when his parents came quietly to bed, though he did not announce the fact. He listened with pleasure to the clicking that was his father's belt buckle, the hushing of his mother's dress as she shook it out, the brief watery sounds from the bathroom. Soon the room was still except for the two other breathings. Jerry opened his eyes. The new moon with a star caught by it was silver in the dark blue summer sky. Then Jerry was by the open window, looking out on the street where the lamplight showed a silent cat crossing the road. The shrubs and trees were so dark a green Jerry could not tell where their shadows began. There was a thick shadow on Miss Jardine's porch, though no tree could be there. The shadow moved. Jerry saw that Miss Jardine stood there with her arms tight around a man. The night air felt warm. Then the shadow split. The man went down the steps. A waft of cigar smoke floated up to Jerry, and a little glowing red thing arced through the air on to Jerry's lawn.

Feeling that Mrs. Grainger would be glad of a visit from them after her time away, Jerry and Mike went to see her as soon as Jerry was well. Jerry drank his lemonade and told her all about Mr. Grainger throwing away his cigar stump.

"He did it before too, in our back yard. Can I play with your green elephants?"

Mrs. Grainger did not answer. She did not look as usual. Perhaps the flu? Suddenly she walked out of the kitchen. The front door opened and closed. Mike went over to the elephants. Jerry did not mind, for here was his chance to investigate the little door in the wall. He pulled over a kitchen chair to stand on and with difficulty undid the hook. The ironing-board fell out and hit him on the head, and Jerry began to cry.

Just A Moment

She told me the most extraordinary story.

You won't like this, coming from me. I'm not the kind of woman you write about. All those women in your stories—long dark hair the lot of them. Elspeth, Louise, old Mrs. Reilly, even that Mrs. Bessemer you couldn't get published. Never thought of that, did you? Look at me. Blonde hair, short smooth cap. And I'm thin. Not one of your warm generous wide-hipped bodies, never a problem squeezing the slacks up over *my* hips. And I wear pale colours. The other day a man at work told me he was having his apartment done in "pale neutrals." Proud, he was. Big deal. I've worn them for years. Beige, grey, tan, cream, oyster, winter white, biscuit, fawn, caramel, oatmeal, sand, vanilla, the lot. Another thing. I'm neat. You don't like that, it's repressed, it's compulsive, better to be messy, *creative*. Well, my kitchen looks as though I'm away on vacation. Nothing on my counters. No pretty jars with pasta in them, no toaster, all in the cupboards where they belong. And I wear make-up in just the way you don't like, so *restrained*. A thin foundation layer, a dusting of powder, I have it blended at The Bay, and transparent lip gloss. Sometimes nail polish, pale pink like those little shells you find on the Islands. Nothing *you'd* notice.

So, blonde and thin and neat and forty-five. I don't like it. Yes, I know, accept our aging and enjoy it and come to terms with it and all that. I don't. Forty-five, it's neither here nor there.

Yes, I still come here regularly. Getting in a rut, aren't I? Not opening myself to new experience? I liked it here the very first time. Back then this decor wasn't fashionable; Maria and Peter just fixed it to remind them of home. Now this look is trendy, plants and light walls and white furniture. Remember when all the fancy restaurants in Vancouver had that awful crimson flock wallpaper, dark wood chairs, hardly see the menu? There'll be some other trend five years from now. But The Olive Tree will look just the same. I always sit here. I have a perfect view of the door so I can—what d'you call it?—observe the passing throng. Yes, it's quiet, even when it's full; we don't get a noisy crowd. Neighbourhood regulars, like me. Of course I've known them all for years. Good solid food, no rabbity salad bar. Good service.

Why *would* I say anything about the wine? You'd just start picking, pick pick, take control of your life, identify what makes you drink, what gap, inadequacy. I have news for you. There's no gap. I like drinking. The taste of the wine, what it does for me. And since you're so nosy I'll tell you more. A *significant* moment in my life took place here. Peter serves a good big glass, and that's what I had for years, four or five good mouthfuls before my meal came, then with, and a couple after the plate was clear. I always eat what I order, no point wasting, can't hold it in the fridge after all. Then I began to realize the glass wasn't quite lasting. Felt skimpy. I hesitated. And on the last night of my vacation I had *two* glasses. On impulse. They were fine. Just fine. I've had two ever since. And something else. There's times I've thought, "I'd like a carafe." One day. And you know, that'll be fine too.

The story. All in good time. Yes, I said extraordinary, it is, but I have to give you the *context*. I want you to appreciate this story, because I'm giving it to you. You're the writer. It's a present. When I heard it, I thought "That'd be fine for her." Because it doesn't mean anything to me.

Back when Ed and I were married and living in Toronto—what? I know I never talk about then. Don't get excited, I'm not telling you " the story of my marriage." Ed just comes into this a bit. He was selling for a stationery company down on Dundas and I was in accounting at Macmillans. Summer. Dreadful August heat. Impossible to keep clean. Always sticky, especially at night. Horrid to be married. So I went to meet Ed at the Bondmaster office after work. We were going to Eaton's to price chairs. I wasn't feeling that good about meeting him. No, no more, that's just a detail you need. I had a new dress, it suited me perfectly. Cream, with a sand stripe, vertical, two-piece, a belted jacket over a sundress. Cool as walnut ice-cream. Cream slingbacks, blonde hair. Of course I don't still have it, why keep a dress twenty years? Yes, that's how long. I was reading a magazine in the reception area, waiting. The chair felt like liquid glue.

The door from the elevator opened and in came a little girl and a woman. The woman was wearing my dress. She didn't look like me at all. Short and plump, bulgy over the belt, ruined the line, and red hair. No, at least not carroty, dark auburn, but all flyaway. Jacket unbuttoned. Worried. Awkward. The little girl? Not like her mother, except for the hair colour. Scrawny. Five perhaps, six, I don't know kids. Never wanted them, never had them. Been crying, I could see that.

So this woman came up to the receptionist, in my dress, and said, "Is Mr. Bracegirdle here?" Imagine going through life with a handle like that! And the receptionist said "Mr. Who?" as well she might, she was almost laughing, and the woman said, louder, "I'm Mrs. Bracegirdle, I'm to meet him." And the receptionist said, "Madam, there's no one by that name here." You know how they can say things, as if you were from Mars. And the woman said—she was getting all excited—"But he must be, he gave me this address, please check." The little girl got going, "Where's Daddy? Isn't Daddy here?" And damned if the woman didn't start towards the offices, past the receptionist, who of course got up and started No no-ing, and right then the door opened and there was Ed.

You know he didn't so much as say Hello? I was right there. Not so much as a nod. No. He looked at the woman in my dress and the little girl, crying, and he said, "Ma'am, is there some way I can help you?" You could see relief going through her. That bass voice. Always sounded so solid, back in Hamilton when we were both in the choir. Didn't mean a damn thing. No, I won't. The woman began singing the Bracegirdle song and Ed interrupted. "Is he by any chance with Stewart Securities?" "Oh yes," she cried. She did, I mean you read that and wonder, but she did really cry it. "That's one floor up," said Ed, "I'll just call and ascertain"—he loved to impress people with words like that—"that he's there." And by golly he did, and he was, and a minute later there was Ed walking this woman and the little girl out to the elevator, reassuring, calming, patting on the head, all that. Funny, I can still see that little girl. She was sniffing, hanging back, as if she wasn't sure it was all copacetic. Maybe just upset at having such a dope for a mother. Couldn't even hit the right elevator button.

When Ed came back into the Bondmaster reception area he looked at me and he laughed. I could have killed him. He always did think I made too much fuss about clothes. No, I don't think she even realized. Oh, she looked at me a couple of times, but not really to see me.

So. Mrs. Bracegirdle and her daughter. No, that's not the story; I said, I'm giving you the background. Have some more wine and I'll get on to the real thing. I heard it right here, two nights ago. What? Well Ed and I broke up, you know that. Not long after. Nothing to do with this. Quite separate.

Two days ago I was down at The Bay. That summer dress sale. Pretty, isn't it? Fresh-looking. This loose style looks nice when you're thin. No, a blend, I can't be bothered with all that cotton cold-water-wash-and-hang-dry. There's one woman ahead of me at the checkout, and the cashier takes my dress and lays it out on the counter—but it isn't mine. The woman in front's bought the same one. Right then the cashier hands over the Baycard and says, "Thank you, Mrs. Bracegirdle."

I tell you, it was strange.

She turned. I must have gasped. Oh, not a moment of doubt. Changed, of course. Her hair's dyed for one thing, not very well, those reds are hard to match I think. She's fat. There's no other word. Too much make-up for the flesh on her face. Turquoise, black, blush. She must take an eighteen, even a twenty. I'd not have thought they'd make that style so big. Of course I never even look at that end of the rack. Loose is one thing, sloppy's another.

Yes, she spoke first. Of course I wouldn't have. What's to say? "I saw you once twenty years ago and I didn't much like what I saw then"? I'm not like you, I don't get all hot and bothered over patterns in life, all that. The second I saw her at Bondmaster I knew we hadn't a thing in common; we still haven't. No. I'd have paid for my dress and gone to the parkade and driven home.

She said, "Don't I know you?"

She waited while I paid—no charge card for me, thank you, I don't donate to department stores. She couldn't remember, but she knew there was something. Her eyes were all searching and wet-looking, as if I were some kind of mystery. I don't like that. So I told her, thinking she'd say "Oh yes" and after a bit of chitchat I'd get free. But oh no. Not enough for Mrs. B. She wanted details, the name of Ed's company, the floor. And then she said, "I remember you, sitting by the window, but not clearly. What were you wearing?"

I told her.

No, not exactly laugh. She said, "That's funny." Then suddenly, "Have dinner with me tonight?"

My mistake: I didn't say No instantly. Startled. Couldn't get the words out. And yes, she'd have insisted on meeting some time; here I'd at least get a meal I like out of it. She didn't give a damn where, I could see, McDonalds or the William Tell. She wanted *me* for dinner. No, that's not right. She wanted to tell me the story.

No, she didn't wear the dress to dinner. Neither did I, couldn't take the chance. Fine fool I'd have looked. She actually wore something that suited her. I didn't say I liked it. She had her hair in a big bun, and in the restaurant light the colour was much truer. A long dark blue robe, a caftan really. Pounds of necklaces, beads, bracelets. Heavy earrings. Her ear-piercings have got all big. Altogether just this side of messy. *Exotic*. The colour worked well on her, though, I'll say that; she has that pale redhead complexion and her skin is still good.

Yes. The story. Here, I'll fill your glass. Don't be silly. I suppose I was just a bit nervous. Before we'd ordered or the wine had come or anything I said, "What's happened to your daughter? She must be grown up now." Well. I'd given her the opportunity she needed. She took the whole damn evening to tell me.

It seems that after all Mr. Bracegirdle *wasn't* up there in the Stewart Securities office. Yes, she called him that all through the story. He must have decided in seconds, because by the time Mrs. B and the little girl got up there he was nowhere to be found. I should think the fire escape.

No, no specific details about their marital problems. I knew you'd ask; well, I'm so sorry I can't satisfy your curiosity. Mrs. B was not specific. She made vague statements, she sighed. So I know, generally: other women. He'd been having a bit on the side. As they say. He'd gone off that morning swearing he was through with the marriage, and she'd phoned him at work and pleaded—she said that, can you imagine? "I pleaded with him." He'd agreed to meet after work. The little girl? Pressure I suppose, how can you do this to your daughter, all that.

So when Mr. B got the phone call from this total stranger at Bondmaster Stationery, bingo, off he went into the sunset. She's never seen him since.

Of course she looked. Haven't you got it clear yet? No pride. She'll do anything. For the next twelve years she looked for him. She'd get word that he'd settled in Uxbridge and she'd haul off with the kid in tow, find work as a sales clerk or a typist, and start looking. Up and down in front of buildings she thought he was working or living in. With the child. Imagine the sight she must have been. Yes, Uxbridge, and Brockville, and Wiarton. Those for sure. Galt, or maybe it was Guelph. And Windsor. Several years in Windsor. I suppose so, a new school almost every year. Sally, her name was Sally.

Funny how Mrs. B talked about all those places. Southern Ontario dumps, the lot of them, but you'd think she'd spanned the globe. Stuff about scenery and buildings and views from apartment windows and trees along the streets in autumn and the Gothic steeples or whatever, all mixed in with Sally.

Mrs. B was crazy about Sally. Yes, but more than most. She wrote poems about her. Talks about her like—a saint, or something. Sally was *it*. Good in school, sweet-natured, funny, lots of friends. Well, maybe she was quick at making friends. The important thing was the marks in school. A lifetime of sales clerking and such—she wanted better for Sally. Prizes, awards, Sally got 'em all. She was especially good at languages, picked them up easy as pie and did a whole pile of them in high school. So Mrs. B had it clear. Sally would study languages at university and would end up translating for the UN for all I know, travelling round the world speaking this and that.

Then one day this little thing happened. Mrs. B kept saying that: "It's little but it's big. A little thing can change a life." Ten times, minimum. She'd had some wine by then, mind you, her voice was husky, and she'd lean across right where you are now with her big fat breasts bulging up over her plate and she'd say, "Just a little thing. Just in a moment."

They were living in Toronto, been there a couple of years. Maybe Mrs. B had finally got the message. Winter, very cold, and Sally went on a school field trip to the Museum. Waste of time if you ask me. They just missed their bus when they left, so the teacher walked the class up Avenue Road to a coffee bar for a hot drink. The kids all fanned out round the room, quite a crowd, and there was someone already at the table Sally went to with her friends. A guy. Twenty-one. Yes, I suppose a big difference at that age. She happened to sit next to him. So—maybe eight minutes in this restaurant. Ten. Not more. Well. You know what Sally told him? Liam, Irish, what a pair for hooking up to men with weird names. That she was a foster child and didn't know who her real parents were.

In the short term it's not surprising, sly these teenagers are, smooth as cream topside for the parents and God knows what underneath. They'd got each other's phone numbers. And he knew about the mean foster mother who wouldn't let her date, who'd whip Sally to bits if she found out about Liam. Well. When a mother doesn't get home till five-thirty or six, and a teenager gets out of school at three—plenty of time for hanky-panky, and that's all they think of these days. Why yes. He was at the chiropractors' college down there on Bloor. He came from out Stratford way.

The thing of it is, Sally kept the secret for five years.

Let's have some more wine.

She finished high school, and just as Mrs. B had dreamed she won a big scholarship to study languages. University fees paid, books, spending money—she just had to keep her grades up. Mrs. B was in seventh heaven. Four years, and a ticket to the future for Sally. And she did those four years, Sally did, and she got first class honours every time and kept the scholarship all through. And she saw Liam. Every minute she could. He'd graduated by then and was setting up his practice in Stratford, but he came to the city on weekends. In the summers Sally told her mother she was going to the Festival with people she knew from the university, and so she got to know his family and friends. And in the spring of her graduating year, when Mrs. B was in hospital for a hysterectomy, didn't Miss Sally go up to Stratford for that whole week to trot round looking for just the right apartment for when they got married?

Of course she didn't get asked to the wedding. The mean foster mother who'd just wanted the government money, kept Sally short of allowance and clothes, even food—why would they ask her? The morning of the wedding, the very *morning*, Sally laid it all out to her. She was going to marry Liam and be a chiropractor's receptionist in Stratford until she got pregnant. She even said, "Thank heavens, now we can stop using birth control." Can you imagine. Then she'd be a full-time wife and mother. No French, no Italian, no Russian. Mrs. B would be welcome in their home for a week in the summer and another at New Year's. *Period.* Then? Then Sally took the train for Stratford and her wedding.

I have no idea what kind of wedding dress she had.

Yes, she tried, but not as long or as hard as with Mr. Bracegirdle. Tired, I guess. Older. Oh, she phoned, and she wrote, and she went up to Stratford. She tried to see them, and she talked to the neighbours and to patients at Liam's office, and she even got hold of Liam's parents. They were sorry for her because of this delusion she had about being Sally's real mother, thought perhaps she wasn't quite as bad as Sally'd made out. They send birthday cards to her, things from the garden, stuff like that. But Sally's never recognized her. Not even when they're alone. Goes deaf. Won't answer. The children call her Mrs. B.

Why is she here? That doesn't make any sense either. She never intended to be. She won a raffle. Doesn't even remember where she bought the ticket, maybe a kid at the door. "A little thing." All expenses paid, a week at the Bayshore, spending money. She thought, "Why not?" A wonderful time—I got her off Sally for a bit and she told me all the places she's been in the past five days. She's seen things in this city I never have, she's talked to people and gawked and oohed and aahed and taken photographs and I don't know what all. Fine, till she met me.

Let's finish the bottle. I never like to leave a bit, it looks so mingy.

Oh, lots of questions. Mostly did I remember anything special about Sally. "Anything at all," she kept saying. Looking for explanations of course. Well, I didn't. I don't. She was a little kid with her mother and Ed thought they deserved his attention more than I did.

I hope you're satisfied. We had a big fight. Yes, then. After he'd sent the Bracegirdles up to their happy reunion. All the Bondmaster staff left and we sat in that wretched waiting area and argued. Never got to Eaton's. He wanted me to stop working, have kids. "Didn't you see her?" he kept saying, "didn't you see the way she was clinging to her mother? Wouldn't you like a little thing running about you like that?" No. I'd told him so before we married. He'd no right to bring it up. He said he thought I'd change. Why do people think people change? They don't. Maybe clothes, that's all. There's Mrs. B now, born to travel and doesn't know it. I know myself. So I left. Yes, you could say pretty well right away. I'd cashed my paycheque and I'd got some extra out of the bank as well, for the chairs, so I just walked out of the office and down to the Union Station and booked on the first train out here to the Coast. The Transcontinental. It left at nine.

About eight years later, when he divorced me. His lawyer tracked me down. I suppose he wanted to get married again. Sure I hope he's been happy. Why wouldn't I?

Of course I didn't tell Mrs. B. Just get her curious, nosy questions, no thanks. You're the writer, aren't you? You're the one with the interest in human nature, the passions, all that? I'm giving the story to you.

To See On the Page

The summer is exceptionally hot this year. In the evenings, after hours of sun, Vancouver lies dazed and radiant with blue heat pouring off the beaches. Walking along Denman Street last week, licking ice-cream cones, my companion and I were among a throng of tanned humans in light bright clothing that moved in columns alongside two slow-moving columns of headlights. Gas, perfume, sweat, coffee, dust, exhaust, ripe fruit and wilting flowers filled the summer dusk. Words and phrases from a dozen conversations floated past us as we walked, not speaking much, aware of nearness in ways not yet limited by knowledge; I played with ways to describe this tension. At a crowded bookstore near the beach, we stopped to flick through the brilliant covers of the magazines and laugh at cartoons. As I took up *Fuse* I felt someone gazing at me urgently, and turning saw a blondly handsome man of about forty who smiled at me with pleasure.

"Ruth?" he asked eagerly.

"No," I said, feeling that surge of annoyance at the idea that one looks like somebody else.

His disappointment was sharp. "I can't believe how much you look like her."

"Is she nice?" I began to turn away.

"Oh—" and he smiled oddly. We moved slowly out to the hot street.

I did not sleep well that night. I was alone; I had thought I probably would be, but there are sometimes surprises. At around three I woke fully from a long drowse. Listening to the rhythm of the all-night sprinkler on my neighbour's lawn and watching the dark curtains sway against a paler darkness, I felt realization work its way through to the front of my head. I knew that blond man. I knew his first name: Carl. He had been a student in a first-year literature class I had taught years before, as part of a night-school program. And I knew Ruth, I knew her too, and I understood why on seeing me Carl spoke her name, not mine.

That class was the kind that teachers dream of. Composed mostly of working people over thirty, the group was highly motivated, experienced, responsive. Lying there in the cool dark with the cat purring by my side (a shorter purr each time because he was going back to sleep), I thought of red-haired Gloria, a legal secretary in her early forties who planned to be out of law school by her fiftieth birthday. We talked once about Anne of Green Gables' conviction that pink was a colour from which redheads were "forever debarred." Gloria agreed: "Too attention-getting." Her papers were models of control, but Margaret Laurence's Hagar undid Gloria and once she wept in class. Sam did so several times. "Hills Like White Elephants" hurt him badly. He had been born and raised in that landscape and the story's dry despair slit the heart of his love of country. At the final exam, Paula and Gordie held hands; she was left-handed, he was right, and there they sat, methodically filling the pages of their bluebooks. At nineteen, they were the babies of the class, the more so because they actually had a baby, which Gordie's mother looked after so they could go to school. And Harold, with the big head that had almost killed his mother, so she told him, who had unpredictable fits of jet-black laughter that nauseated many of the students; and Anna, perturbed always by any suggestion of sexuality, who went crimson when she realized "Lay Your Sleeping Head, My Love" was about two men; and sad Nancy; and Pauline, who had left a battering husband years before and now ran a successful bottleparty business—she had no patience with "these women's groups"; and Greg the tax accountant, who argued every point. . . . All were generous beyond measure to a new teacher.

For reasons that are no longer clear to me, I had organized the material we read according to theme. We therefore "did" a group of stories and poems about love, which generated intense discussions about the roles socially prescribed for women and men. Then we did a group about what the old syllabi called "Man and Nature," and finally, suitably, we did a group about death. Besides *The Stone Angel*, we read Faulkner and Porter and Purdy and Atwood, Plath and Dickinson and Thomas and Yeats. Harold—of course—did a report on Mitford's *American Way of Death*. Feasting on literary death, we listened to recordings, tapes, our own readings; I think we spent part of an evening looking at photographs of totem poles and death masks and English brasses.

One foul December evening, after we had struggled and dazzled our way through Plath's "Two Views of a Cadaver Room" and Auden's "Musée des Beaux Arts," I could tell that everyone was truly tired. So was I. There was Eliot yet to go. I did not think I could take the class through it. Now, I would simply open the remaining hour for discussion, knowing that worthwhile insights would emerge; then, I felt that we must at least stick to our topic, and so proposed that each student describe a personal experience related to our reading.

I hope that class realized that their young teacher did not appreciate what she was asking for.

Stories of parental deathbeds. Childhood pets (Gordie and Paula). Young friends killed decades before by the dreaded August polio. A colleague healthily at work in September, fluish in October, dead of pancreatic cancer by the New Year (Greg). A nephew laughing at high school graduation and two hours later dead, drunk, on the Upper Levels highway. A husband dead of a cerebral hemorrhage, leaving a young wife eight months pregnant—that was Josephine, the oldest in our class, a grandmother now and years into a happy second marriage, whose voice shook as she said, "It still gets to me that he never saw our Gretchen." A rejected lover, a suicide statistic three months later, and all the guilt. The gentle grief at the death of grandparents, not ill, just through with living.

There remained two students, Carl and Ruth, to tell their stories.

"Mine's a bit different," Carl said. "In mine, the death didn't happen. It just almost did." He smiled, emphasizing that this story wasn't going to be a wringer; many students smiled back, for Carl was likeable, warm, talkative, the kind who responsibly contributes to discussion. A dentist, he'd taken "quick looks" at many mouths over the term, and once did a class demonstration of The Right Way To Brush Your Teeth. I still do it that way. How could I not have recognized him?

Early in his practice, Carl had had a patient who needed to have all four wisdom teeth out. They were impacted, infected, a mess. Carl arranged for the anaesthetist who worked for many of the dentists in his building to come in and, seeing his patient well under, began his work. "It was all perfectly ordinary," Carl said, with his pleasant blond-moustached smile, ordinary until the Muzak system began playing "Sweet Georgia Brown." "I can't stand that song," Carl said, laughing, "and I looked over at the anaesthetist; I think maybe I wanted to see if he hated it too." The man was gazing out of the window, abstracted, unaware of the patient—who, as Carl then observed, was scarcely breathing and whose skin was a colour he had never seen before.

"Oh, another thirty seconds or so," Carl responded to a question. "Brain damage at the very least, if not—." He stopped, trying to order his thoughts. "It was all so random. I mean, if they'd played 'Some Enchanted Evening' instead, the guy'd have died. No, I don't know what the anaesthetist was thinking about. But that's why I don't do anything that requires anaesthesia any more. I send 'em upstairs. Let someone else handle it." He smiled again. People sighed, and moved in their chairs. Then we all focussed on Ruth.

Such attention was new for this student. She was the kind whom I used to see as filler in a classroom. Such students, neither gifted nor lacking, and having no special appearance or personality, were simply there, their work on time and of the assigned length, their grades in the Cs. They had a hard time with poetry. Ruth was a thin, dark woman of about thirty; her midlength brown hair was drawn back with pretty tortoiseshell pins. She leaned her head to one side and stroked one of the pins almost constantly, perhaps as an aid to concentration. She was a housewife with children, I knew, and her views were unremarkable. Like many others in the class she had preferred Yeats to Thomas, Laurence to Atwood.

"I am dying," Ruth said.

"Oh, not immediately," she added into the silence of the classroom, and she spoke with a quirk of a disappearing smile as she looked down at her desk and stroked her hairpin. "March or April I should think."

Josephine got up into the silence and started over to her, but Ruth held up her hand. "You don't need to do that." Josephine sat down again. "I'm used to it. Anyway"—her eyes slid round to me—"this is just an assignment."

Everyone looked at me. I knew Ruth sensed that I did not want her to continue.

In her monotonous voice, she said, "It's been strange reading all this stuff about death. I thought at first, 'Well, this will be funny.' A lot of it isn't right. They mix things up too much, saying that life and death are one, all that. They aren't. One's one, one's the other." Here Ruth began to straighten up in her chair as if becoming accustomed to being the speaker, the centre. She looked directly at one student after another. "I liked 'A Bird in the House,' though." Again her expressionless eyes came round to me. "What you call the foreshadowing. All the things I didn't realize were clues." She looked at Carl. "I'm the patient when they played 'Some Enchanted Evening.' Oh yes, I could have been fine now, if they'd caught it."

Now Ruth was upright. She stopped fiddling with her hairpin and folded her hands calmly. "I could tell you all about how they didn't catch it, but those details—they don't matter. I guess that's why I like the poems better in this course. They get to the point faster." Again the sliding eyes. "Even though you didn't give me that good of a mark on my poetry paper." Again the little quirk of a smile. "But a lot of them are silly too. Easeful death and he kindly stopped and rage rage. That, with chemotherapy."

I could hear other classes down the hallway breaking up.

"I'll say one thing for this course"—the eyes again—"it's got me into reading. And I've found something that says what I think. I've heard some of you say that," and Ruth glanced about at Paula, Sam, Gloria, Anna, "and I thought you were putting it on, this great feeling, to see on the page something you felt but didn't know how to say." She picked up her class notebook. "I've got it here. Don't worry, it won't take long to read." She stopped flipping through the pages and I saw her mouth begin to move.

"Here. 'I have seen all the works that are done under the sun; and behold, all is vanity and vexation of spirit.'" She paused, frowning. "There's a better bit, shorter—here. 'Vanity of vanities, saith the preacher; all is vanity.'" Taking hold of the two pretty tortoise-shell pins, she lifted her wig off her bald scurfy head. Looking right at me, Ruth said clearly, "Class is over now?"

Here memory compelled me to get out of bed; lying still in the semi-dark was intolerable, intolerable to hear the cold hissing of the sprinkler on the black grass. I pawed over the chair by the typewriter and found a jacket. Slipping it on over my nightgown, I began to walk.

I walked round and round among the rooms of my house, blundering over furniture, handling and stroking the strangeness of unseen bookcases and tables and cushions. Although the carpet and the bare floor were alternately frictional and cool beneath my feet, although I tried to walk firmly (I am a big woman), all the while I felt as if I moved through clouds of tatters and confetti and rags and tags of on-the-way-to-nothingness. Fabrics, substances, textures shredded and frayed and undid themselves and blew away, spiralled in the force of an unfelt wind, spiralled away, away. Then I began to go. There was no pain. There was only the sensation of loss, of advancing nothingness, of absence, as they all dislimned, all the events and experiences of the long years since Ruth's bald head. The arrival of two children. The departure of a husband. The endeavours to write. The political struggles. The manoeuvres to retain a job. All that effort, all that immense effort invested in the maintenance of my own existence turned thin, volatile, less than air, oh much less, much. My parents, my family, those monstrous figures clustered at the back of the stage, became as nothing. The new figure in the foreground, the man I hoped would become my lover, smiled his sweet crooked smile; for a second I felt his long skinny arms round me; then he was nothing, and then there was almost nothing left of me.

After the long night I saw the typewriter on the desk, the litter of files and papers, the rough drafts, notes, outlines, revisions, inserts. All that.

Imperatives

There was no pen in her schoolbag. There was no pen. But surely she had put it there that morning. She could see the surface of her table in her room at home; there was no pen on it. So it must, it must be in her schoolbag. It wasn't.

Nicola hunched stiff in her seat and stared at the inkwell. Which yesterday she had somehow tilted out of its wooden hollow. Which had rolled, blurted ink all over notebook, tunic, floor. Which now held once more its dark blue pool. And from which she must now fill her pen as she had been taught, and watch the endless amazing blue silk unreel at her hand's bidding as she copied neighbour, flavour, reindeer, painful, reign, aging from the blackboard. But there was no pen in her schoolbag.

Half the class was absent with flu, and already she saw Sister Michael moving down the next aisle but one as she checked each girl's work. The nun's glasses shone with a hard light and her broad face was pale. She had been very angry about the ink. And about the lost pens. Nicola's stomach pain started. If only she were back with pencils. Even Sister Michael didn't mind much if you lost pencils. Certainly not Mum and Dad. But she didn't really want to be back there; the day you went from pencils to fountain pen was almost like going into a new grade.

She and her father had chosen the first pen (cardinal red her mother said it was) just a few days before baby Jimmy was born. Later in November she had lost it, and her parents were almost as sorry as she was. The blue one went early in January. This new school in Toronto required a lot of equipment—zippered folder for art, schoolbag, pencil-case, linen bag for gymsuit and shoes—and she could never remember in which classroom she was supposed to have which container. Somewhere in transit the bluebird had gone. And now this green one, brilliant green, though her mother had not said it was like a parrot or anything. Brand new two days ago. And her father had looked at her with his tired gentle eyes and said, "Honey, you just can't go on losing pens like this. You're big now, soon you'll be eight. And you know money's tight with us right now. So if you lose this one, Nicola, you'll just have to get along without."

Now here was Sister Michael.

The terrifying questions.

The whispered answers.

The ruling: "Nicola, you will do no more writing in this school until you have a pen."

The great black triangle moved on up the aisle. Her soapy serge odour filled Nicola's nostrils and the touch of her large hand was still cold on Nicola's wrist. Other girls glanced at her covertly. She stared at the grooved scarred blotted woodgrain of her desk. Heat in her skull and cold in her shoes. She could not come back to school. She stared at this fact, rigid in her seat, until the bell rang and she could go out in the snowy day and walk through the little park and so home.

Baby Jimmy was crying. He had something called colic and always cried. Hour after hour, hour after hour, all day. At night too. Nicola woke hearing him, hearing her parents' low voices, hearing the rhythm of her father's steps up and down the long hallway of the apartment. Her mother did not walk the baby because she was not well. She was not *sick*; her father had impressed that upon her, to counteract her fright at seeing the great red blotches on Mum's nightgown a while after Mum and Jimmy came home from hospital, and the terrible upset of Mum going back there not once but twice with something called haemorrhage. But she was *not* sick. Sometimes this happened after a baby was born, and soon Mum would feel better if she got lots of rest, and Jimmy would stop having colic, and then. But still Jimmy cried, and Mum lay in bed with her eyes closed. Nicola could often tell that she was not asleep.

She had no pen. She could not ask for another one. She could not go to school any more.

"How do you like your baby brother?" her parents' friends asked. Nicola did not know what she felt about Jimmy. Once she had seen her father holding the wailing baby and shaking him, shaking him, and Dad's face looked like a Halloween mask she'd had once, the mouth stretched wide and angry and the eyes bulging. And once she had seen her mother nursing Jimmy and crying. The tears did not make any noise. They slid from under her closed lids and fell splat on her rumpled nightgown. On both occasions Nicola went away fast and quietly. Sometimes she could hold Jimmy for a few minutes. He felt nice, small and warm and squirming. She did not know.

What could she do? What could she do?

Two things had not changed since they moved from Vancouver to Toronto, since the new school, since Jimmy.

Every morning Nicola's mother combed her hair for her, and braided it, and put on the tortoise-shell-look barrettes with the elastics underneath. The comb ran smoothly through her hair and her mother's warm fingers nibbled at the back of her neck, and then there was the quick rhythm of the three strands binding together. "I love you, Nicola," she said, and kissed the child on the forehead. Then she lay back again and Nicola went to school.

Every evening Nicola's father read to her, and Jimmy's crying faded as Dad's voice moved through *The Wind In The Willows*. Sometimes he even laughed, at Toad, and that made Nicola feel happy. They talked. He knew she was almost making friends with some of the girls in her class, and that she did not like Sister Michael much. She knew why she was going to this strange school, so unlike the public one she'd attended in Vancouver—the nearest public here was seven blocks away, across two heavily-trafficked streets. (Nicola also knew, but not from direct telling, that her parents had to pay a lot of money for her school now.) He knew that she enjoyed arithmetic and that they were doing balance-beam work in gym. She knew that when Dad finished his "thesis" and had his "degree" many things would change and be as they had been before or maybe even better. Then he sighed. "Well, I've got to get back to work, honey. Time for bed." Tonight, lying stiff under blue flannelette sheets, she heard the tap-tap of the typewriter mixed with Jimmy's crying. She got up and searched for her pen, even looking in silly places like her jewel-box which she hadn't opened since Christmas.

She could not tell them.

Eventually she cried, deep in the pillow so they would not hear her, and eventually she slept; once she thought she heard her mother's voice.

"She looked odd. Upset or something."

"I just checked on her," said her father. "Sleeping peacefully. Go back to bed."

"I wish I knew more about that school. When I'm better. . . ."

In the morning Nicola got up and washed and dressed and ate the oatmeal her father cooked for her, with the nuggets of wheat germ and brown sugar hidden at the bottom of the bowl. Her mother braided her hair, she kissed Jimmy and her mother kissed her. She could tell her mother wanted to see her eyes, and turned quickly to get her schoolbag.

"Have a good morning, dear," and then she was out on the porch and walking down the steps. Then down the sidewalk, and then around the corner and on her way to—but she could not go to school.

And then she knew what she was going to do.

The park lay before her. An ordinary Toronto park, with a dead fountain in the middle and snowy mounds which were summer flower beds, and gravel walks, and an area where the swings and seesaws would be after the twenty-fourth of May. Around the edges of the park were benches. She would stay here. She would stay in the park till lunchtime, and then go home, and then stay in the park again until three. And nobody would know.

Time went very fast at first. It was strange, delightful, to be alone in the white park. No crying Jimmy, no fierce Sister Michael, no tired father and in-bed mother, no couples and trios of little girls who all went to each others' birthday parties and couldn't be asked home to play because things were not right, though they would be soon. The trees in the park were lovely, white and black, lovely to look up to and count and feel. She took off her mitten and stroked the wet rough bark. Much of the snow was untrodden, and she made patterns on it and sang and talked to herself. She circled the flower beds and scraped ice off the sealed fountain, kicked the snow off the hopscotch pattern and hopped about. It was very quiet. The few cars going by made a low murmur as their wheels revolved in the soft whiteness, and only occasionally one or two people walked through the park. The bells in the nearby church tower regularly cast their clear tones into the air. There was one great pine tree, so dark a green it was almost black.

Sister Michael had succumbed to flu early in the morning, and lay in the nuns' little infirmary with a high fever. Young Sister Laura had taken over her class on five minutes' notice; she had no instructions. There were more children absent than present, and she did not think to make an enquiry about Nicola Gordon.

Time slowed down suddenly when Nicola began to feel cold. Her winter coat and stockings were thick and her boots waterproof, but they did not add up to the warmth of a snowsuit; this was her first Eastern winter. She set herself to run round the block formed by the park, and she was comfortable again for a while. There was a man sitting on a bench on the side opposite to her base. He was smoking and looking at the snow on the ground. When she ran round again half an hour later he was just throwing a match away. He smiled at her but she did not smile back because she was not to talk to strangers.

The bells chimed eleven-fifteen. Three-quarters of an hour until she could go home. She was very cold. She needed to go to the bathroom; the second she realized this her need was terrific. But she could not go here. The only cover was a clump of bare-branched shrubs, thin and spindly. No. She would have to hold on. She tried sitting still on her bench with her knees tight, and counted the ribs of her black stockings to take her mind off her bladder. This was hard. Eleven-thirty. The man was coming towards the middle of the park where the four paths converged at the fountain. She sat very still. Her feet felt icy, and her stomach hurt. He walked round the fountain and back the way he had come and out of the park. Relief, relief. She stumbled up and walked again, first all the way round the park and then criss-cross on the paths. As she finished the third tour, eleven-forty-five rang out. Soon, soon, soon, soon, she hummed as she made three more tours. Now surely she could start for home; no, she must wait until the first chime of noon. She stood in the snow and listened so hard her ears hurt, thought she heard the air vibrate, took a step—but she was wrong, there was no bell. When it did come she was so dazed she did not believe it, and the third bong was soaring through the cold air before she broke and ran all the way home.

"My goodness but your cheeks are rosy! Did you run all the way home?"

Nicola nodded as she rushed past her mother to the bathroom. Lunch was a grilled cheese sandwich, oh hot, hot, wonderful. And applesauce, still warm. How this? Dad didn't know how to make applesauce. She looked at her mother.

"Mum! You're dressed!"

Nicola's mother smiled, looking down at her slacks and shirt.

"I'm feeling a bit stronger, Nicola. Finally. So I got up and got dressed and made that. And Jimmy hasn't cried for almost two hours this morning. Good, eh? There he is now."

Nicola scrubbed her eyes with her napkin when her mother left the kitchen to get the baby.

She delayed leaving until her mother said, "Nicola, is something wrong? Is there . . ."

"No, no, everything's fine. See you at three. I'm going to run all the way again." And out the door and down the street and round the corner and into the park; and there on her bench was the man.

"Hello," he said, and smiled at her. She'd run so fast she was almost on top of him, and she just couldn't not smile back.

"Come out to play some more, eh? Don't you get cold?"

"No," said Nicola quickly, "I like the cold."

She ran off to the hopscotch and played there a while, and then worked slowly round to a tree near her bench and took a good look at the man. He did not look bad. He seemed a bit older than her father and, like him, very tired. The hair showing from under his hat looked faded; his topcoat was too big for him. He was smoking another cigarette and staring at the little black spot in the snow where the match had fallen. His left hand was in his coat pocket and the right one was ungloved.

Nicola walked up to him and said, "Hello."

"Hi there. Looks like you're pretty good at hopscotch."

She smiled, and stood silent looking at him.

"How come you're not at school? I guess you go to that girls' school down there, eh? Seen the uniform before."

Nicola's face shut.

"Trouble, eh?"

Nothing.

"You live near here?"

"Yes," she whispered, ready to run.

"Well, then," and he sighed. The bells chimed: one-thirty.

"How'd you like to build a snowman?" he asked.

"Fine!" said Nicola.

He did not fuss, even though his hands were bare and the knees of his trousers got soaked and it took quite a long time. They agreed that a snowman should be made of three not two balls of snow, and though a hat was not essential a nose was. They spent a while finding a stick of the right conical shape, and gravel-lumps big enough for eyes, nose, buttons. They admired their work then, and a couple of passers-by smiled at them.

"He should be holding something," said Nicola.

"Yeah, I know, but what?" They looked around.

"Hey!" said the man, and held up a last-summer's popsicle stick.

"I know!" and Nicola ran to the pine tree and scuffed through the snow and found a pine cone. "Here." The man slit the bottom of the cone with his dirty thumbnail and Nicola slid the popsicle stick in.

"Here, honey, you put it in his hand." There was the snowman holding an all-day sucker, and they laughed.

The bells chimed two-forty-five, and they sat down on the bench near their snowman. The man lit a cigarette and coughed loudly.

"Looks good," he said cheerfully.

"Maybe I'll bring him a scarf tomorrow," said Nicola. She should not have said that. The man's mouth was opening and he was looking at her and she blurted, "Don't you go to a job? My dad doesn't, he's a student, a special kind, he's called a graduate student. But I thought men went to jobs?"

"They do," said the man, "when they have them. I haven't got a job right now. Not right now. Graduate student, eh?"

"Will you have one soon?"

"Sure I will. Sure." He smiled at her.

"Are you going home at three o'clock too?"

"No, not right then. I'll wait a bit, walk maybe. Get home a bit after five. Looks like we're going to have some more snow, eh? Sky's all thick."

Nicola and the man sat companionably and looked at the dour grey sky. The man took her mittened hand and held it. She did not think this was supposed to happen and was not sure if it felt wrong or not. When the bells called three she got up.

"Home to your mum then, off you go."

"Are you going home to yours?"

The man laughed the kind of adult laugh that Nicola did not like. "No. Wish I was! Home to my wife." He threw away his cigarette. "G'bye honey."

Sister Laura closed the door of the classroom with relief, exhausted after the day's unfamiliar work. She went to the staff room and called the infirmary. Sister Michael seemed slightly better; her fever was going down. Yes, Sister Laura could come in the morning, during first period when the girls were in gym, and discuss the day's lessons with her.

Nicola opened the apartment door to hear her father saying, "I've told you and told you not to take it too fast. You shouldn't have stayed up so long. Now back to bed with you. God knows we don't want this to start all over again." Jimmy was crying. Nicola kept out of the way until suppertime and things seemed better then. Her parents joked and said maybe they should rename the baby Screamy or Hollerer. Later her mother washed her hair, which she usually loved having done, and her father read her a complete chapter of *Wind* instead of the usual five pages. But the snowman was huge in her mind, and the man, and not having been to school. Her stomach hurt.

In the morning Nicola's lunchbox stood on the kitchen table beside her bowl of oatmeal.

"You'll stay at school for lunch today, Nicola," said her father, and patted her head. "You've done that before, right? You know where the lunchroom is and everything?"

"But why do I have to?"

"Your mum's got doctors' appointments today, for her and for Jimmy. Check-ups. Come on now Nicola, you're not scared are you?"

"No."

Nicola went to the bathroom twice before leaving home, and squeezed hard to get as much pee out as she possibly could. When she got outside snow was falling, falling in the steady determined way that she knew meant it was going to go right on doing it for a long time. Her bench in the park was thick with snow. She sat down anyway, with her lunchbox and her schoolbag on her lap.

Then she saw him. The man came hurrying across the park, hurrying as though afraid of being late. He sat down and put his arm round her shoulders and looked fierce, but his voice was gentle.

"Listen honey. I know you don't want to talk but I have to know some things. We can't stay outside today. Fine for our snowman but not for us. How about going to school?"

Silence.

"Then will you go home to your mum?"

"She's going out. I have my lunch. I couldn't anyway."

"Tell me why?"

She shook her head. The man took his arm away and leaned back against the bench and sighed.

"I brought him a scarf," said Nicola, and went over to the snowman and tied it round his fat neck.

"Okay," said the man, "okay. Let's go. We'll go ride the streetcars and look at the stores, and I'll get you back here at three."

Sister Michael spread out her class's attendance sheet on the rough woolly infirmary blanket.

"Has Nicola Gordon got the flu now too?"

A few minutes later Sister Laura came scarlet-faced into the staff room and sat down by the telephone.

The Bloor car sang through the falling snow. Nicola and the man sat at the back, eating raisins from her lunchbox.

Sister Laura put down the phone after ten rings. Then she began looking through the child's file. Wasn't the father a university student or something? There must be a department office. . . .

End of the line, end of the line. Lovely words. Now back all the way to Yonge, and a transfer to another streetcar, and then down to the department stores. The man told her about the subway that would open in a couple of years, and she liked the idea of it; but surely there could be nothing better than the swaying smoothness of the trolley, with the snow fluttering against the windows and the warm leather smell of the seats.

The English department secretary took some time to establish that Geoffrey Gordon was at that hour scheduled to be in a seminar on Renaissance poetry, that he was not there, and that he had gone to the library.

The man wanted some coffee and a smoke when they got to Eaton's, so they went into the huge fragrant warm cafeteria. Nicola poured her soup into her thermos cup and then took off her coat. She glanced up and found that the man was looking at her in a way she knew meant, "What a pretty little girl." She did not like it. She put her coat back on and started to drink Dad's tomato special.

Sister Laura finally made Nicola's father understand that nobody knew where his daughter had been yesterday and nobody knew where she was now. A brief and horrible silence passed between them. All he said then was, "I'll get a taxi." Bang. Sister Laura began to whimper. Sister Michael came shakily into the staff room and said, "Rubbish. Call the police and get them here before he comes."

It was strange, delightful, to be in Eaton's toy department at a not-Christmas time. There was leftover Christmas candy on sale. The man bought Nicola an all-day sucker with silver foil wrapping; there was a red-and-white Santa printed on it. She put it in her lunchbox to save for later. Then they went over to Simpson's.

A police car picked up Nicola's mother and Jimmy at the clinic, and meanwhile her father left the school and went to search the apartment. After he had gone through all rooms, all closets, he did ridiculous things like looking under the toaster. On his way back to the school he saw the snowman. He arrived in the school staff room bearing Nicola's scarf.

Jimmy was crying. Nicola's mother took the scarf and tied it into a knot and untied it and tied it again.

Nicola could not imagine anything more cold and grey and endless than Lake Ontario with the snow disappearing into it. Then they walked over to Union Station, and Nicola shared her sandwiches with the man; he had said he was not hungry, but she noticed that although he had finished a pack of cigarettes he did not buy any more. They each ate half a peanut butter and half a marmalade. She wished he would not sit so close to her.

"She didn't have her pen," said Sister Michael.
"She didn't have her pen?"

As they waited at the streetcar stop the man said, "Listen honey. Whatever it is, it can't be as bad as going on like this. You tell 'em. They'll understand. People do, you know."

Radios, alerts, messages, descriptions. A school photograph for the police and one for the newspapers. Sister Michael said, "The responsibility is mine," pushed Sister Laura aside and got into the police cruiser with the Gordons. Jimmy lay quietly now on his father's lap, blinking at the snow.

The park was utterly white and silent.
"Thank you for the all-day sucker," said Nicola, licking.
"That's fine honey." He looked at her worriedly. "Now you understand, eh? What I said? And. And listen. If you come tomorrow I'll, I'll have to go to your school and tell."
Nicola could see he hadn't wanted to say that. She made herself nod, and turned her face up to him as he bent to kiss her, and the black car silently drew up alongside.
When the black car pulled away again the man was in the back seat. That was the worst. It was worse than seeing Sister Michael and her father cry when, finally, holding her sucker tight and looking at Jimmy's round eyes and feeling her stomach going hard-soft hard-soft hard-soft, she said, "But you told me I couldn't be in school without a pen and you told me I couldn't have another one." Much worse.
They could not get her to let go of the sucker. She refused, she refused, and when she went to bed that night she still clutched the bright sticky packet in her hands.

Twoscore and Five

We were all then in our twenties.

Jack was a painter. So was Marjorie. At the back of their dark small Kits apartment was an unexpected door into a north-lit studio where they really lived; nothing blocked their view of the North Shore mountains. They had not even a houseplant on the sills. Jack was "macho" (the term was not then in use). Present or absent, he dominated conversation with Marjorie. His asthma required attention. Marjorie talked, cooked, cared, smoked. He painted far more than she. The point is this: she knew. To her women friends she spoke openly of her efforts ("struggle" was not used then either) to believe in her own work's importance and to act on that belief. Her preference, which ran counter to received radical artistic opinion of the day, was for small canvases. She would not marry Jack. Well. While I was away in library school in Toronto she left him and took a studio of her own. Her postcard with this news pleased me; many of her friends were pleased. In my next letter I told her how I looked forward to hearing of the progress of her work. Would she plan a show soon? But she did little painting in the new workplace. I had not expected that.

Then she suddenly left Vancouver, with Leo, whom we had all known in university. He had been close to forty then, glamorous with age, poems in print, critical articles, affairs with students and junior faculty. "Leo's latest" was a sought-after position. What is the desire to be defined thus? He was now almost fifty. His poetry was in my opinion undistinguished.

Nobody knew his wife; evidently Marjorie was her last straw. The house was sold, the children distributed, and Leo was off to his new job at U of T, with Marjorie. She turned out to be a poet. We remembered then, discussing the event: she had written poems in college, off and on, and she had always given her paintings unusual titles. Formed as lengthy phrases, they were suggested routes for the viewer to travel into the work. They returned to the mind.

She and Leo had rented a lovely old house in the Annex. I was living as inexpensively as possible, in one room, in order to save even small amounts towards my garden and my house, and found it most pleasant to talk with Marjorie in her turret study—"Yes, it's really a room of one's own"—and to look out over the spreading, abundant foliage of the beautifully mature maples and oaks that border the streets there. Marjorie and Leo did little with their garden. Admittedly the exposure was difficult. Marjorie said herself that to work was not easy for her. Although the child was Leo's, Marjorie was the adult at home when Jake got out of school. Leo was the established writer, she the roadie; considerable entertaining was required to establish him socially at the university.

Leo lasted seven years, the same as Jack. I had been back on the coast five years by then, and although I was still working at the Collingwood branch I was about to transfer downtown, to the Main branch. My savings were accumulating. I viewed houses in Kitsilano and Dunbar regularly, having particular blocks of particular streets in mind for the best possible combinations of terrain and outlook. A southern exposure was essential. Marjorie wrote that the worst part was leaving the boy. They had apparently become close. Leo placed severe restrictions on her contact with Jake. She thought Leo resented her refusal to marry him. It is certain that he did nothing for her as a writer, in spite of all his connections. In spite, I suppose. Only a handful of poems had appeared in little magazines (I checked the new issues regularly as they came into the Literature section). She created some striking images. One poem presented a train journey east from Toronto, in winter. Glancing up from her newspaper, the speaker of the poem saw at the grey lake's edge, just in from the stony beach where the white ice-sludge crashed and waved, an apple-tree, bare-limbed of course, but still carrying half a dozen frozen apples. They were no longer scarlet, but blackish, like dried blood.

When I wrote in response to the news of Marjorie's departure from the Annex house I praised this poem, expressing the hope that her new situation would permit the writing of many more of such quality. I got a note back almost at once. I had not expected that. Customarily, weeks and even months passed without a word from her. Then would come a long, long letter, its various parts written on different dates and kinds of paper with different writing implements, its content so miscellaneous (recipes, work in progress, Jake, the English department, jokes, weather reports, comments on the biographies she was always reading) that I found it as difficult as it was absorbing to read. Or postcards would come, mailed on successive days and forming a serial message like the old Burma-Shave commercials, or there would be a beautiful card of paper handmade by an artist friend. Often her writing not only covered the designated areas of the page but also ran clear over whatever printed message there might be, or into the space intended for the address. Sometimes Marjorie even wrote on the backs of envelopes. I myself wrote to her once a fortnight, regardless, on regular bond and the typewriter—or, in the last year or so of her time in Toronto, on my computer.

This time Marjorie had painted a card herself: a beautiful watercolour illustration of a skunk cabbage. They were thick in the marshy parts of Stanley Park just then; by chance I had walked there only the previous day, and their strange colouring, gamboge and brilliant green, was sharp in my mind. Here it was again. Such accurate rendering of detail is not often found outside specialized botanical publications. Marjorie wrote that she had found a design job with a firm of medical publishers and planned to write on her own time. She added, "Dear M, your steady letters and steady love help keep me going. Have a drink! Start smoking! Love as ever, Marjorie."

That year was the first Christmas I went to Bermuda. People at work thought I was going for sunshine. In a sense that was accurate, for my desire was to observe the vegetation of another climate, the fruits of another context of soil and atmosphere. I wished to see that other life, to see for example the indigenous Bermuda cedar that I knew only through photographs; it is really a juniper, and nearly died out earlier in this century. Stopping over in Toronto to visit Marjorie, I found her full of stories about a doctor she had met through her job, who was setting up a community clinic in her neighbourhood. She was painting murals for the reception area and writing leaflets. When I thought subsequently about our conversations, I realized she had evaded my questions about poetry. Instead, she had heard the difficult and dangerous story of the events which had led to my new job in Acquisitions, and learned my timetable for buying my garden and my house. I had not expected to tell her all that. I was not surprised, though, when Marjorie wrote that she and Gordon were living together. Discussing the news with mutual friends, I realized that Marjorie's life was like a story, or, more precisely, a story retold several times in different ways. Some of the major novelists tell and retell essentially the same tale; I do not object to this in fiction, but the clinic ate up every hour Marjorie chose to give it and asked for more, was Gordon's far more than hers, formed the basis of their social life.

He was not an unpleasant person, Gordon. I spent time with them each December in the four years they were together, and found him gentler and kinder by far than the other two. He was no gardener—only a couple of spider plants and an African violet sat dryly on the kitchen windowsill of their apartment over the clinic—but he took at least an amused interest in my plant-smuggling from Bermuda, recommending ways to calm myself physically before the encounters with Customs. When I finally made the purchase of my garden and my house he also made some useful suggestions about shelving; many of my books on gardening and botany and natural history are outsize or otherwise unusually shaped, and so required me to create custom shelf designs, and these in turn needed to complement the type of wood used in their construction. Fiction of course fits easily on to store-bought shelving.

Gordon ended it, though. Marjorie would not have a baby. She had always been clear that she did not want children, first because she was painting and then because she was writing poetry. I do not know what reason she gave Gordon. Her letters did not say. She was apparently in pain. How is it possible to give up a known person for a hypothesis?

After Gordon, Marjorie came back to Vancouver. My promotion to assistant department head had just been confirmed. I felt strong in victory. Marjorie looked ill and was. A Toronto doctor had recommended a hysterectomy—discussing this had generated the break with Gordon—and a Vancouver doctor confirmed that diagnosis.

Marjorie would have the surgery and would then come to my house to recuperate. Unlike our other friends, I could offer her a resting-place with no children, no pets, no men. Of course a very great deal remains to be done in the garden, even at the level of basic design, for I have had just two growing seasons here and am only now becoming familiar with the variations in the soil, the drainage, the disposition of the existing trees and shrubs, the angles of light in the twelve months, the ways the shadows fall from house and garage. This year the new greenhouse and solarium will introduce yet more variables, and although I of course made extensive calculations when designing and locating these additional structures, I have learned, with difficulty, with reluctance, that it is not possible to predict the activity of plant life. What I can only term bizarre capacities are present, for growth, for refusal. Bermuda made me see that first. Perhaps the sight was easier in another climate. To work deeper into such difficult garden seeing is like going from reference to cross-reference, shelf to shelf, text to text, in search of an answer. Extraordinary developments, though, can take place in greenhouses. This Christmas I plan to bring back some begonias. I have tried nothing so ambitious yet. The deception at Customs unnerves me, brings me into risk, the narrow airless corridors at night and the men with shining glasses, but I want to attempt those exquisite Bermudan begonias here, to try my best to anticipate and overcome the difficulties of persuading them to accept as theirs another climate. The delicacy of their colouring is remarkable.

Whatever the external weather, in the solarium there is always light and green. As I told Marjorie before her surgery, she could be in the solarium all day if she chose, comfortable, healing, neither overheated nor damp nor chilled, reading the catalogues and watching the bulbs come up while I did the spring pruning. She wept. "God, Mary, it'll be wonderful. I could use a lifetime of that." That is exactly what I have.

In the meantime I visited her daily in the Vancouver General. She looked small, frail. Grey was coming in at her temples because a tinting was overdue. I thought of all she had done for Jack and Leo and Gordon. For several days Marjorie's pain did not permit conversation. She lay, I sat and held her hand. Then she entered the bored, irritated stage of recovery, unable to abandon herself any longer to hospital passivity and equally unable to read or talk seriously. She ordered me to start talking.

"I can't just lie here, Mary, I'll go nuts. Tell me about your job, or something. What are those things people have? Hobbies."

Well. Although "restraint" brings many dramas to the administration of public libraries, I have never been able to make the intricacies of budgets interesting to friends. So instead I told Marjorie about the risks people take in libraries: the men who expose themselves in the Boys' and Girls' section, the patrons who use X-acto knives to slice sections out of illustrated reference books and take them home, the patron who, before we caught him, went through four shelves of Shakespeare using a fluorescent yellow felt marker on every appearance of "breast." (There are many.) Having told these brief tales, and having established that Marjorie still did not play bridge, I cast about for something more substantial.

"What about your lovers? Tell me about them."

"I don't have lovers, Marjorie." Technically this was not so. I do on occasion share a bed. But that is all I have felt like sharing. She closed her eyes and moved her head restlessly, turning towards the window. The sun shone on her upper body. The weatherbeaten skin of her throat and chest stopped, and below was silken cream. From where I sat, the cream met the yellow of the budding daffodils I had picked for her, and this colour in turn met the blue air over the North Shore mountains.

"Well then, tell me what your life is about, at forty-five. What do you do when you're not alphabetizing or chasing old men out of the Heidi section?"

"Forty-four," I said, and knew. "Marjorie. I'll tell you about—about A and B and C."

A decade ago, an intelligent and highly-qualified woman I worked with lost a promotion to a foolish and underqualified man. She and I never thought of the library the same way again; we had not understood that merit was relatively insignificant in the matter, and had waged no campaign. We have since more than regained that lost ground. At that time, I developed a link with a women's organization in the city. To energetic activists, keeping periodicals and clippings and books and pamphlets in order (alphabetical and otherwise) is not an appealing task, but some of them do recognize it as essential. Twice a month I go into the office, more often if there has been a flurry of feminist action in Vancouver. I feel odd in that office, yet I like it. The young women are friendly, although or perhaps because I do not appear to be like them. I listen to their conversations as I work among the files and shelves. Doing so, I am reminded of browsing; I rarely browse now, knowing the classification system as well as I do, but many library patrons of course do, dipping into a page here or there to sound out a book, to know if they want to enter further. I like to watch them. Sometimes I bring flowers, which the young women always welcome. Most of them do not habitually observe the natural world, absorbed as they are in their interactions with each other and with various opposed and allied organizations. That is why I only bring cut flowers.

The story I now planned to tell Marjorie I had learned through overhearing at the women's office; I had heard it in instalments, as it were, over the year or so the narrative had taken to play itself out.

"What kind of story? I didn't know you were a libber, Mary." Marjorie turned from the window and looked at me critically.

"Political, really."

She closed her eyes again.

"It isn't boring, Marjorie. It's about three young women."

"Oh Mary, you're not going to call them A and B and C, are you? If you must tell me this tale then I'll think of names for the characters."

Impatiently I waited, because the more I thought about my story the more I felt it was perfect for Marjorie now. Jobless, manless, babyless, stripped of routines and entertainments as of her own always distinctive clothing, my old and dear friend could now observe her own naked lineaments and design her own life, not one predicated on someone else's design for his. Marjorie could grasp that certain emotions are insubstantial bases for decision-making; that if you do not act in your own interests, others will impose theirs on you; that—

"Alicia, Beatrice, Caroline."

"Why such old-fashioned names?"

"Because mine is. I like it, and I've spent years defending it against people who want to call me Marge. So start, Mary."

The previous spring, the women's organization had decided to publish a monthly magazine and for this task had initiated the formation of a collective. Of this, Beatrice and Caroline were members and Alicia was not; she did fund-raising work for the organization as a whole. Shortly before the first issue appeared, the collective discovered that the woman who had agreed to solicit advertising had not done so.

"Actually what they said was that she had fucked up."

"I've never liked that expression," said Marjorie crossly. "It makes fucking equivalent to a mistake. Go on, Mary. This isn't interesting so far."

The collective gratefully accepted Alicia's offer to do a one-shot job of selling ads. She did so, very well, and came to a collective meeting to report and hand over her files. When she left, Beatrice rose and excoriated Alicia for having usurped the functions of the woman who had originally assumed the task.

"But—"

"Yes, obviously. Everyone reminded Beatrice of what the process had been. But Beatrice said she had been talking to the woman—"

"What was her name?"

"Oh Marjorie, surely we can just call her D."

"Daphne."

"If we must. Beatrice said many things that began with 'If Daphne were here I know she'd say. . . .' She would not stop. Then Caroline spoke up, defending Alicia. Beatrice became angrier still. The argument drew in the whole meeting. The women never reached the rest of their agenda."

"There was something else going on," said Marjorie, "there must have been. All this emotion." She opened her eyes.

"How did you know? Beatrice and Alicia had been lovers previously, and now Caroline and Alicia were lovers."

"Complicated." Marjorie giggled, and winced. I did not see why she giggled; there had also been giggles when I had first heard this part of the story. I had felt uneasily that such hearing was an invasion of privacy, but neither the teller nor the other listeners had apparently even been aware of me sitting next to them. Perhaps they assumed I would not comprehend, even not hear. Many professionals assume their conversations are incomprehensible to the layperson.

"To continue. Beatrice and Caroline worked together in the magazine collective, arguing all the time. I should explain that all three women lived in the same communal house, along with other members of the organization. The collective used the basement to do the magazine layout."

"Even more complicated. What did these women look like, Mary?"

"I never met any of them, Marjorie."

"Oh well. I'll just use my imagination."

Why, when the obvious question was this: how could anyone by choice enter such a situation? But I went on. The magazine was well-received. Beatrice and Caroline fought; they had full opportunity to do so, for Beatrice was deeply involved in the production process and Caroline as deeply on the editorial side. Alicia and Caroline were happy.

Then came the Thanksgiving weekend, for which the lovers planned a trip to Seattle, where they would stay at the home of an absent friend, attend a Holly Near concert, browse through the University bookstore, eat Mexican food.

"A little honeymoon," said Marjorie approvingly.

"I suppose. When they were about to leave—Alicia was in the car, with the engine running, waiting for Caroline—Beatrice came rushing out and asked, no, pleaded, to stay at the house in Seattle with them. Another woman's trip had fallen through, and Beatrice had a chance at her air ticket and concert seat, but she could not possibly afford a hotel. And Alicia—"

"Said Yes without asking Caroline!" Marjorie burst in, half-rising on her elbow.

She was right, again. But again I could not give Marjorie the detail she wanted: How had Caroline reacted? What had the weekend in Seattle been like for each of the women? I knew only that after "a lot of hassle" Alicia had assumed the mantle of peacemaker. ("What an operator!" murmured Marjorie.) She urged that when they all got back to Vancouver Beatrice and Caroline should meet formally, perhaps even with a mediator, and reach an understanding. "For the sake of the magazine as well as for their own sakes," I added, repeating the phrasing as it had come to me.

"Yes, well." Marjorie sniffed. "Do get on with it, Mary. Did they meet?"

"They made a date, some weeks away—"

"Who was stalling?"

"I am sure I do not know, Marjorie. They were both very busy. You have no idea how active these young women are. They go to meetings constantly."

"Boring."

"On the day of the dinner Caroline developed the flu. It was November, raining of course, and cold. But she was not able to reach Beatrice at work and cancel, so she dressed and got herself out to the agreed-upon restaurant."

"Beatrice never showed." Marjorie smiled, and scrabbled in the bedside table for her cigarettes.

"You shouldn't do that."

"I know I shouldn't." She lit up. "This is the first one that's tasted good since the op."

Difficulties were developing with the magazine. The springing vigour of the first few issues had metamorphosed into a regular production cycle, wherein financial and editorial problems grew monthly larger. Beatrice and Caroline disagreed yet more sharply and frequently. ("Well of course they did," Marjorie said with irritation.) At the New Year, the collective did a self-evaluation. The analysis was that more members were needed.

"Alicia!" Marjorie waved her cigarette.

"Well—yes. She had done hours of volunteer work for the magazine, and that was a recognized way to become eligible. But Beatrice said she could not work with Alicia under any circumstances. She refused."

"And she reminded everyone about poor Daphne." Marjorie inhaled with delight.

"Marjorie, how can I tell this story if you keep on trying to tell it?"

"I'm just telling you more about it while you tell it. What then? What reasons did Beatrice give?"

"She said that Alicia was 'too bourgeois.' A quite incorrect use of the term; they use political terminology so carelessly—oh, very well. But that made no sense, Marjorie. There were other well-off women in the group and Beatrice had never objected to them."

"They hadn't taken away her lover."

The meeting failed to resolve the issue. Everyone left for home exhausted.

Over the next few days Caroline phoned round to various group members to say that in the interests of the collective she thought both she and Beatrice should leave. An emergency meeting took place. Would the collective not expand but shrink, and at that lose two of its most hard-working members? Caroline recanted. Believing that she did not bear the major responsibility for the dissension, and that she did have a useful contribution to make, she would remain. Beatrice said, "I resign, on political grounds," and walked out.

"Oh dear." Marjorie thoughtfully stroked the sheet over her incision. "Poor thing. She still felt she had to put it that way. And Madame Alicia was waiting in the wings?"

"Well, yes. She came right in with the other new collective members. They proceeded with the agenda. There was such relief, you see."

When Alicia and Caroline got home that night Beatrice had moved out of the communal house. She left her job next morning and the women's community in Vancouver saw her no more. Someone said she had gone back to Alberta.

Shortly after this the organization, concerned at the way the magazine ate money, decided to pull back to a simple gestetnered newsletter. The collective dissolved itself without argument. Alicia and Caroline moved forward to other projects. A new group formed, for the small newsletter task, and when the letter from Beatrice came no one at first understood what it was about. Who was this woman on the prairies who claimed the collective had slandered her, and put about a rumour that she'd been expelled? She wanted a retraction, in print. The new group couldn't see the point. They put Beatrice's letter—it was quite long—in the "Hold" file. When the long-distance calls came, various women in turn said they were really sorry but there was no one there who could help.

Marjorie sighed. The sound indicated both sadness and satisfaction. "So it was Beatrice who made the magazine be."

"Marjorie! She wrecked everything."

"Oh Mary, that's not how it was, you've told it all wrong." Marjorie sat right up, wincing and looking stronger than since her return to Vancouver. "It's a love story, Mary, you must see that, it's about the power of love to move people, change things. Oh, Beatrice must have been very much in love to be so strong. Going to Seattle. The dinner incident—that got her more time, of course. Fighting with Caroline so as to stay bonded somehow to Alicia. Keeping even a little separation between the two lovers by fending Alicia off from the collective. And living in the same house! Probably she had to overhear them making love. Oh, such determination, such a fight for her love, don't you see? It's quite clear. Beatrice's emotion was the power. The magazine lived on the tension she created. Of course her leaving meant collapse."

Marjorie lay back smiling on her pillows and turned away from me to the window, where pigeons flapped in the brightness. The mountains were brilliant, the snow a harsh dazzle. What to say? Haphazardly I visualized the possibilities for my rock garden: alyssum of course, sedum, potentilla, aubretia in its subtle shades, snow-on-the-mountain, perhaps cotoneaster or another prostrate. Attempting such growth is risky because just there, in the only place where the rockery can be, the soil tends to clay and will need careful treatment before planting, close attention after. I know already that some of the plants will fail to thrive, for reasons I will not be entirely sure of, and similarly some will do extremely well.

"God, it's good to see those mountains again," said Marjorie. Her voice shook. I had not expected that. "She didn't quite get out in time, did she. The letter, the phone calls. Poor Beatrice. Over the edge a bit. You have to learn to get yourself out in one piece." She looked again at me and showed the tears. "I know her, Mary. I've not been in that exact situation but I've felt it all happen. When did Alicia and Caroline break up?"

"Did I say they broke up?"

"Well of course they did, without Beatrice."

"As a matter of fact it was just recently, when Beatrice was telephoning."

Marjorie took another cigarette. I did not speak. She blew smoke reflectively. "Soon it'll all start again for them." She seemed neither sad nor happy in saying so.

"Surely not Beatrice and Alicia?"

"Oh no no, not that specific pairing. The pattern."

I considered this. In university we had had to write essays concerning novels and plays. "Do you mean that in Beatrice's life there will always be an Alicia and a Caroline, and that in Alicia's—"

Marjorie wagged her cigarette in agreement. "Yes, that's right. You do see, Mary, really?"

I was pleased. "A and B and C." Perhaps even now I could make my point.

"Oh Mary, what will you do with the rest of your life?"

"I was about to ask you the same thing."

"No, you were going to tell me what to do, Mary. I can feel what you think of my life, my dear, it's all in how you hold yourself. So stiff in the chair there. Well. Probably quite soon I'll fall in love. Oh no, I'm not over Gordon. I never really get over them. Not Jack or Leo either."

"They are compost, perhaps." I had not meant to speak aloud, but Marjorie reached for my hand and gave her lovely smile, unaffected by the lines and the altered texture of her face.

"Yes Mary, I can feel that I'm ready to love again. A man wandered in here yesterday looking for a patient who'd just been discharged, and we got chatting for a while. I wouldn't be surprised if he came back to visit me. He's an architect."

"You're certainly making your way through the professions." I tried to take my hand away.

"Mary, do try to understand. I've got so much from each one, lived in such different ways—I've almost been three different people. Four, counting me. You've misunderstood me, you know. I don't have it in me to be a great poet or painter or anything like that. I have a bit of talent, enough to give me and a few others some pleasure, that's all. It isn't really what I'm about." Marjorie stubbed out her cigarette and clasped my fingers now with both hands. "Oh Mary, I wish so much you'd take a chance on a person. Garden, books, bridge. You know what you should do, Mary?"

"What should I do?" Marjorie seemed unaware of the sarcasm. Her eyes focussed brightly on her idea.

I thought of the vegetation in Bermuda, how fantastical and strange it first seemed to me, so luxuriant, enveloping; but I went back, again, again, and came to know those plants, travelled to some familiarity with the casuarinas, the bougainvillea, the lovely trailing vines and the myriad begonias. I do not truly know them. I doubt even the possibility of real knowledge. But I rely, yes, I rely upon that annual visitation to another country.

Marjorie was bursting. "Look out the room door, and the next man that comes along, go after him and say, 'You're so handsome, I just had to speak to you. Come have a drink.' Look, let's see who comes." Laughing, she peered round me to the hospital corridor. For half a minute no one passed. Then came a tall youth, perhaps eighteen, carrying freesias. He strode by and out of sight.

"There he is, Mary!"

"Marjorie, hush, for heaven's sake. He's probably visiting his girlfriend."

"No, his mother or an aunt. The wrapping's from the hospital flower shop. For his girlfriend he'd have gone to a real florist, and got bigger flowers too. Go on!"

"Marjorie, freesias do not need to be any bigger than they are." I turned from the door and clasped my other hand over hers. Marjorie leaned back again, flushed, relaxed, alive with smiling humour, and we looked at one another, two middle-aged women friends, each gazing at the mystery of the other's life.

Porridge and Silver

"Oh I should never have said that!" cried old Mrs. Reilly and woke violently. Her heart knocked, tears accelerated into her hair. "Oh God, not again." Half past three.

She glared into the darkness, gave a long shuddering sigh—I sound like a car on a corduroy road—and began the special breathing. In nose, out mouth, in nose, out mouth, barely bend a candle-flame. *Stiff* corduroy, Dad's knee, a tiny car sliding down his thigh. Gently now. It's hard to be gentle enough. Her younger daughter Jane had learned the breathing in prenatal classes, to which Mrs. Reilly had accompanied her several times. Disturbing, the notion of a husband at a birth—but she had learned a lot. Dad laughed a lot. I didn't understand why he was so angry when I took the corduroy road out of the hamper and into the sandbox. He found out. He always did. Amazing what I didn't know, and I had three. In nose, out mouth. Small sounds, up and down in the dark. Much later he laughed again, not at, I think, with. Too late, I was hurt. Long enough? She checked her heart. Heavy regular thumps, gradually softening. Mrs. Reilly sighed again. On to asphalt now. Out here it would be a logging road.

With extreme care, she initiated small flexing motions in her knees. Bad tonight. Do I have to get up? Check. Yes—but if I'd slept through to seven-thirty I'd have held it. Odd. Would it go all the way back to toilet-training, that absolute connection between waking and urinating? Mother always said she was so proud because I was "clean" at eighteen months. That's—seventy-two years ago, heaven help me. From whence cometh my help? And the bookroom at one-thirty, I must get some sleep. The dream-misery seized her again, she twitched, she cried out as the arthritis bit and her bladder distended. "I must get up," said Mrs. Reilly angrily into the dark, and wiped her eyes. Doing so hurt her hands.

She reached down until her wedding ring rubbed against the piping of the little bolster and pushed, which hurt, and simultaneously moved her torso to the left, which also hurt. A good one. But why do I remember only the bad? dream the bad? Forty-nine years of a good marriage, it was so, why does only the bad come up in me? Another push, move. Right foot just over the edge. Another. Foot dangled, knee twinged. Mrs. Reilly grinned because she knew the pain was peaking. The spanking he gave me, hit hit *hit* HIT and then all over. Soon downhill. Amazing, how motion works out the pain. Cold engine. Couldn't talk to him about it, of course, children didn't—what do Tom and Jane call it, discuss, yes—we didn't discuss with our parents then. One more push and also move, left foot well over too, that's good. Now the bad. The crooked elbow. All the things I did, said, decided wrong. Mostly, I remember the early years. Early in the morning. What does Elizabeth call it, obsessing, obsessed with the wrongs I did. Him. Us. Them. The elbow. Now.

She slowly bent her right arm so as to lean on its elbow and thus raise her torso off the bed. This hurt so that she hurried, always swore she wouldn't and always did, leaned hard and thrust down hard and so came up hard, back straight and legs swinging over the edge. Dizziness boiled into her head. No I will not close my eyes. Twenty to four. Before her was a chair, and from its back depended her cane and dressing-gown.

Yes I will say it. "There's no one to see me. Not any more," said old Mrs. Reilly sharply to the chair, as if it had been in error. I could go padding around in my nightgown, or without it for that matter. But *I* would know. Slopping about. This next bit isn't too bad. Oh I should never have said that. "Move to Point Grey? So you want to start getting mentioned in the social columns, is that it? No thanks, I like it here." Oh I can hear my young voice, sarcastic, contemptuous. I didn't want to know what it meant to him. Reach over to the back of the chair; slide; slowly, woman! Remember Tom and Elizabeth and Jane skidding down the slide so fast, squealing, laughing? How we all loved that garden on the slope. He tried for years to train the roses up the side of the garage, but they just tilted and rambled and swung where they would. Yes, but he got that tree-peony to bloom. Years—he moved it three times—and then cascades, springs, waterfalls of peony blossom. Dayspring. And all my beans tomatoes cucumbers carrots beets lettuce. The asparagus bush, he loved that, lacy, graceful. Every year the pumpkins for the children at Hallowe'en. Oh stop wandering. Dressing-gown. From on high.

Grasp back of chair firmly with left hand. The gown is folded so the opening to the right sleeve is uppermost; slide right hand and arm in, instructions in a car manual, then hold chair with right hand, and use left to push right sleeve up to catch over right shoulder. Tricky; but I've not fallen yet. Now lean to the left, yes ouch but it's peaking, there, just till waving clutching hand can find hole in slithery stuff and get *in*. Where is it, now? There. Straighten up; heart, slow down; in nose, out mouth. Better. Now right hand, be my right hand, go over the breast, the breasts, my breasts, still lovely if I say so myself and no one else ever will that's certain, death and taxes, unless when they lay me out, cream-silk skin, go, connect with the left lapel. Pull. I'll droop my left shoulder to help you. There. Done it again.

Again. Done, not done, done.

Yes but why, tell me why? Me, tell me why. After Elizabeth was born I healed badly. I couldn't tell him. I should have made myself. Then one time I cried out. He looked. He *looked*. He said gently, "Why didn't you tell me? You don't tell me things." He made me go back to the doctor. Said he'd phone him, *phone* him, if I didn't. Minor repairs, I went in for minor repairs. I wonder if that's why Elizabeth is a psychiatric nurse. How tender he was, after. But still I could never talk to him about it. No. I wouldn't, though I knew he wanted me to. Oh dreadful of me. My knees still hurt dreadfully. Non-conformist knees. Meekly kneeling's not for me. But the rest is all right now. Mrs. Reilly grasped her strong cane, rod and staff, and began to walk slowly out of the dark box of her room.

Yes, but I didn't want to move. I loved the big frame house on the Slope, the windows, oh the windows, north to False Creek and the mountains. From whence. Yes, but still—Mrs. Reilly struck the carpet with her cane—I should have listened, tried to understand. I'd had it easy all my life. He hadn't. And that house he wanted to buy didn't have two steps up from the sidewalk to the walk and nine from the walk to the front porch and we wouldn't have had to move when he had to go into the wheel chair. He always hated that apartment we went into from the Slope. And that house that he wanted was brick, like the house I grew up in back east. Odd, I never thought him a hating man. He was so good to me. Yes, but maybe not for me. Don't. I didn't mind that apartment.

Mrs. Reilly negotiated the dark turn into the little hallway, and did not bang her hip on the dresser. Left-hand Louisa he called me, always had fits driving with me in heavy traffic. Yes, but it made him laugh too. He needed that. I think I didn't mind the apartment because I knew I wasn't going to die from there. Not that you ever *know*, the day of the Lord so cometh as a thief in the night, I haven't thought of that for years, but what I mean is, he knew he *was* going to die from there. Or at least go from there into the hospital, same thing, one-way street. I was just going into a smaller space. I liked that funny tiny kitchen, the drawers sliding in and out like silk, the dishwasher. The big old kitchen on the Slope never looked so clean no matter what I did. It was, though. This one's even smaller. Selfish, selfish. Smallest one yet to come, not yet I hope, why do I hope?

Yes, but back then I wasn't thinking about steps and wheel chairs, why ever would I? I thought it was that "I want to be recognized as successful" thing in him. Couldn't see why he needed that, still can't. Yes you can, don't tell lies. You didn't, so you wouldn't see that he might. He didn't start out from where you did. Mrs. Reilly neared the long mirror which hung near the bathroom door. The first thing we chose, bought, together. Why I wonder, I mean why the first? I can't remember that. He liked it because it was old. Something about other people, past people, having looked into it for all those years. I wanted it because of that nice curving frame. I wonder how I look. Do I look tonight? Yes. Teach you to hurry that elbow part. Punish. Elizabeth would think I'm crazy. So would Jane. Tom mightn't. Well, you're not telling them anyway, are you? Get on with it.

She looked into the dim silveriness of the mirror. I suppose that's why he kept on going to church, too. Neither of us believed that any more, and I wouldn't go—all those years of being *made* to go. The idea of him being a miserable offender! But he said the language kept him coming. There is no rhyme in the English language for the word porridge or the word silver. Strange that we always see ourselves reversed. Fifty years of driving and still sometimes I think That must be a European car behind me. One after another all three of them thought, and searched, and asked their teachers; and no, there isn't. Aren't. The words of the King James. I think Tom feels that too. He goes, at the great feasts.

Yes, but I don't have to linger, do I? Just look. Tall, thin, never that awful dieting so many women think they have to. Have to. Another thing I didn't do right, all those early years. The school skipped Tom into Grade Three and he was so shy, Jane's awful chemistry teacher, the one who laughed at the girls—all those miserable things that happen to children in school—and he'd be away for a day or so on a trip; and I'd go ahead. Saw the teachers. Saw the principal. Did what I thought should be done and didn't discuss it with him first. And then it was too late. Why? Mrs. Reilly hit the carpet with her cane and looked some more.

So I can wear a quilted gown without looking like the side of a barn. In fact that long pale fall of pink looks beautiful. Yes, but my face. Am I that cross always? Old grouch, old witch, short hair though, couldn't brush the long any longer. Jane and Elizabeth swore up and down, why up and down? that it would work, mini-what? Afro. Mother would have had a fit. So would he. And then loved it. No no no don't think of sitting on the bed in that room on the Slope, rain sluicing the windows and the garden and the water's surface and the mountains, how beautiful upon the mountains, and him behind me brushing and brushing and brushing, me warmer softer looser with every stroke, never a word between us, what word would do? there is no rhyme, and his hand sliding over the silk. Stop. Of course he'd find out. What was I trying to hide? That there'd been trouble at school. I'd tell him what I'd had to do about it.

Often, not the right thing. He'd not say so, I'd figure it out from the silences. Shame. Once he said, "You're treating me like a back seat driver in this business of the children and their school. I'm the father, you know." Oh I knew, I knew. Jane and Elizabeth wouldn't understand; they think both parents should do everything. Jane nurses the baby through whooping cough and then Ted takes her to the doctor after, though he's seen nothing of her illness, being away on that trip—where's the sense in that? I did think the children were more my work than his. I never thought his job was mine. But there was something wrong, something I did or didn't do. When Tom had that awful time in Grade Twelve he went to his dad anyway. I didn't hear a word till it was all over. I still don't know exactly what it was all about. Just like the time that idiot rear-ended me and all I thought of was Jane, my baby, was she all right, and there was Tom in the back seat swallowing blood from the cut lip till we got home and out of the car and I saw his face. I couldn't talk to him either. Still. Still, I haven't caused an accident myself in fifty years of driving.

"Aaaah!" said Mrs. Reilly loudly and screwed up her face till her eyes disappeared. "You're an ugly old woman." She went into the bathroom, her bladder feeling hot and tight. The toilet was easy now. Tom, Jane, Elizabeth had ganged up, insisted, phoned workmen, and rails had been installed. Hitch up night- and dressing-gown together, one side at a time, then grasp with both hands, down I go, let go. Which feels good. Yes, but the children. Tom. A souring marriage, I suspect, and the boy with dyslexia. And Jane, why did she have so many? Middle-aged, and her last not yet in school. Only mild dullish pain now, no bites pinches pangs. Going back to my room will be a, what does Elizabeth say, a snap. She makes snaps of too many things. I wonder why she never. All to do in the morning again, but too bad. It is too bad. Wandering again.

Now. I want to know why. Why, when he's dead a year and more and the babies are forty-eight and forty-six and thirty-eight, the afterthought, why when I only have a bit of arthritis and bifocals and high blood pressure and occasional insomnia and I forget things sometimes, mostly that I am old, why is it that all that comes up from the past is bad? And there is no health in me. Things I did wrong, hurts I gave, mistakes I made. When I turned sixty-five he said jokingly, "And you've done no one visible harm." What a backhander! But that's really quite high praise—assuming that all harm manifests itself to sight.

"Wandering old fool." Mrs. Reilly reached for the toilet paper. Whoever put the holder there must have thought orangutans were going to live in this building. She looked at the pink paper wad in her clenched hand. Well, look, woman. Knotted lumpy twists. And the colour of them! So cold. You'd think there hadn't been a drop of fresh blood in there for a week. Well, maybe there hasn't. Maybe the blood just goes past the gates marked Hands (and Feet too for that matter, and she glanced down at the bones cased in their pink fur), just zips right on like a car going past a freeway exit. When I think of all these hands have done. Flexible strong ingenious capable warm, each with four fingers and a thumb to take a fallen eyelash out of a baby's eye, steer a car through blinding rain, twist the cap off a jar of home-made dills. Gone, and here are these—things—instead. I wish it was as simple as eyelashes now. Were. No, that's wrong, I know that's wrong, they don't want that, good children never do. Oh why must I always think wrong? Not think really—feel, and then the feeling turns into thought. How does that happen?

Unable to remember if she had wiped herself, she reached for the toilet paper again. Orangutan. There. That's what I mean. Every time. I don't know why I do that. But I do know I must get up.

Mrs. Reilly did so.

I must get up, so I can go back to bed, so I can get up again. Bookroom at one-thirty; leave before one, weatherman said Occasional Rain and after forty-nine years in Vancouver I know what that means. So nap from twelve-fifteen to twelve-forty-five; so lunch at quarter to twelve; so back from shopping and library at half past eleven; so leave here nine-thirty; so two hours for up, bathroom, dress, breakfast, kitchen, coffee, newspaper, crossword, one cigarette, call a child so they all know I haven't died or broken a hip or gone mad overnight, mail, bathroom, coat, purse, keys, cane, and out. Oh God I wish I was dead. No I don't. Mrs. Reilly avoided the mirror. Were. Yes, but that always sounds so awkward. It is I. It is I.

The small noises of her breathing went up and down in the dark. It is I. In, out. All the children tried it, all children do, to stop breathing, and of course you can't. There is no rhyme. I shouldn't have. Close your eyes, woman.

Another sound began, the continuous obstinate softness of water stroking and stroking and stroking the window and falling away in the dark. The dew of thy blessing. The coast, the coast, I have loved the coast. Mrs. Reilly slept.

The windshield wipers on the old Volvo did their work well, but slowly. There was no wind for the water to play with, and so the heavy rain fell disconsolately straight down in strings like newly-washed hair, and wove silver veils over the old car's hood and over oncoming headlights. The cool damp from her open window freshened Mrs. Reilly's left cheek and made the warm leathery car interior feel even cosier. Compensating for the wipers, she leaned forward like a racing driver. Straight ahead; then check inside rear-view mirror; then outside; then straight ahead again; and watch that traffic light stop sign turn signal large van sports car muscle car and taxi; and a bad camber next intersection but one, remember it in this rain.

"Nothing like it," said Mrs. Reilly, smiling as she curved decisively past a bus. Broadway now, and a clear run to the bookroom. Other widows, she knew, dreamed of standing before bridges whose ends they could not see, of happily boarding trains to unknown destinations; in dreams, she drove her city. Mostly. She put her foot down and enjoyed the firm press/give between the muscles and their accelerator. There's a parking space; tight. Yes, but I don't *want* power steering, I like doing it myself. Oh all right, all right, next time. OK Jane? Tom? Elizabeth? And I'll get it in a bright red diesel Rabbit, too, and won't you be surprised. He'd have a fit. And then love it. I suppose I should say *if* I get another car. Well, unless the arthritis gets a lot worse, I will. Mrs. Reilly saw hood and rear align with the vehicles fore and aft, and relaxed her fingers. Drive—that I can still do.

"But why, Louisa?" far too many of Mrs. Reilly's friends had asked her, when she began her volunteering for the society a month after the funeral. "To meet people, dear?" In a way; usually, soon after they got the news. They were desperate for books, pamphlets, articles, tapes, brochures—words—to tell them the name and nature of this morbid hardening, this debilitating lignification that did its hard scaly work invisibly within and left them piled in wheel chairs with great stinking diapers lapped about their devastated limbs. They wanted, oh how they wanted, words to tell them why. No answer, of course, no answer, no rhyme, Mrs. Reilly knew that, been through it all with him. No, friendships did not grow from these encounters with humans newly-balanced on the verge of despair. Sometimes, intense communion. Sometimes, a sense that her elderly coolness and methodical ways said to them, "There is still order in the universe, in spite of what's just happened to you." Maybe they didn't want to hear that; but there she was. Sometimes, sympathy. This was for Mrs. Reilly a spongy unreliable word, but there were people, especially the young, who did call it out of her. She could feel it sieving through the lines on her face. Often there were tears. To meet people? No, not in the usual sense.

"Well then, are you doing it because of him, dear?" "Because of Dad?" (That was Elizabeth, of course, though the others probably put her up to it.)

The answer that repeatedly formed itself in my head: "It's the dailiness." I couldn't say it, they'd laugh. There was a little room with polished windows, lined with drawers and files which she herself had neatly labelled; there were shelves bearing bright pamphlets and magazines with their black print inside, carefully ranged by date of issue. To these Mrs. Reilly entered, five afternoons a week. Here she never wept or berated herself, never felt anguish pulse acid over heart's tissue at the memory of an act, a sentence of thirty thirty-five forty years before.

The supportive discipline of routine. Breathe in, breathe out. What you do for children. Daily bread, that ancient prayer. Give us this day; stop there, right there, and you'd have yourself a fine prayer. Good hot oatmeal every morning. Be there, reliably, when they get home from school. Or when, five ten years married, they phone you and their voices simply aren't *there* in the words they say. First Tom, then Jane. And why did Elizabeth never? Nor in the bookroom did she think of her silken breasts, her ugly hands and feet. Each day gave mail to be opened, new publications to be displayed, fresh cards to be written—yes, writing hurt, but the script was almost as clear and firm as ever. All must be completed and ready for use once more. Each day, phone calls; some encounters with patients, with the regular staff; and then they hung up or went away, and the quiet order alone remained and the small noises her presence made in it. Late each afternoon, Mrs. Reilly became aware that she was tired. This was a pleasing sensation. She gave herself permission to think, briefly, of sherry and cigarette and dinner at home;

alone; yes, well, even though alone. A spoonful of wheat germ on the bottom, one two three four five, and a spoonful of brown sugar on the top, one two three four five. Each day. Mrs. Reilly still made it for herself in winter sometimes.

The silver rain flowed even more thickly now over the hood, down the car windows, along the gutters gleaming under the lights of cars and shops. Her knees were stiff, all that sitting, she made herself flex them. Twenty more minutes of hard work, good, and I'll be home. The traffic was so thick she could hardly hear the Volvo, but of course it kept moving along, solidly, smoothly, well-maintained as it was. The gear-change sequence repeated, repeated, and the traffic lights went through their metamorphoses. Remember when I found out Tom thought it was me, I, who made the different colours come and go? Mrs. Reilly smiled. Roadside clusters of pedestrians thinned into lines as their umbrellas floated them across the wet streets. There's fog starting. Very slow now. What's this? She peered forward till her nose touched the windshield. Traffic jam. Ah. Clearing now. There we go, and the accelerator gave beneath her foot. A thick twisting grey thing blurred before her eyes and disappeared downwards. Then came a dreadful lift and then a subsiding of the Volvo's left front wheel.

"I have killed a man," said Mrs. Reilly, and she said, "I'll never drive again."

No you haven't they all said, and Yes you will, again and again, witnesses, police, children, witnesses, police, children; and her lawyer said the same thing after the children got in touch with him. That man had stepped right out into traffic, walking quickly, a block from the nearest light or crosswalk, in rush hour. He had not looked where he was going. Not a glance. It was dusk, rain streaming, fog festering. She was in good health. Her bifocalled eyes were fine. The Volvo was a wonderful car, and those wipers are well within tolerance ma'am, don't you worry. There is no fault. You could not help it. He did not look; he was old, ill, a heart condition apparently.

"All coverup. The plain fact is that I've killed another human being. I'll never trust myself behind the wheel of a car again."

Mrs. Reilly was as alone as she could get. She had refused to go for the night to Tom's, Jane's. "I want to be in my own bed." However, Elizabeth's insistent form lay on the living-room sofa. Mrs. Reilly pulled the sheet over her head and murmured into the cotton. "I've killed a man. My hands were on the wheel, my foot was on the accelerator, my eyes saw his shape, my car went over him. I am responsible." Elizabeth had made her swallow pills; exhausted and in pain, she fought against their encroachment. "There's not much time. No visible harm, that's what they're saying, but a man is dead dead dead and I have killed him." Mrs. Reilly went on like this for some time and then lay very still, her face covered. She had forgotten to put the bolster under her knees. The rain streamed down, unheard.

Towards six in the morning the tumultuous forces in Mrs. Reilly's brain overcame the chemical troops deployed in her bloodstream and she awoke, harshly. Vomit? No, just swimmy from that muck. I have killed. I have to go to the bathroom. I can't look in the silver mirror. I can't ever drive a car again. I could get the bottle out of her purse, I saw her put it there, and take them all. She wouldn't wake. My legs don't usually hurt this much. She slept through everything when she was little. I think she did.

The image of her daughter—competent, professional, reserved, administrative—lying starfish in her crib with the moon pulling silver ribbons over her flowered pajamas made Mrs. Reilly cry at last. The tears erupted and rolled and fell, tears for the dead man, for the dead men, for the children's childhood gone, for her old and lonely self with no one who would face, recognize, validate the dreadful thing the dreadful things that she had done. The old man had been what is called "alone in the world." I *have* to go. A wife a daughter a son a sister a brother, I'd go, go on my knees and beg forgiveness, forgive me my most grievous fault, weep with them, help with the arrangements, pay for the headstone. But there aren't any. I can't go in the bed, it would be too awful with Elizabeth here. Oh God my legs hurt. I'll wake her up with all this sobbing if I'm not careful. My heart, my heart. Breakfast will be bad enough. Stop that. No. In nose, out mouth. Mrs. Reilly snuffled and snobbed, thinking of the candle-flame, and in time the frantic beating rediscovered order. She could feel the stiff dried tracks of the tears. I must get up. It'll be cold, November, we'll have oatmeal. and what then? Take the bus to the bookroom? I could. Turn the mirror to the wall, no good, bad, the other side is always there. I *must* go to the bathroom. Well, I could. There's a stop just outside.

The other side is always there; Mrs. Reilly lay still with her legs tight together, hurting hard, and stared through the darkness.

I didn't see him. That's true.

But his being there, his being who he was—that wasn't my doing. That's true too.

And I, it, was what he desired. He chose.

And I drove as well as I knew how. Given the same for me.

This fresh knowledge flowed through the scaly carapace of grief.

"I couldn't be everything. No one can," whispered old Mrs. Reilly, humbly. She thought then of the old and new pains ahead of her, and felt self-pity sneaking up to fiddle with her tear ducts. "No," she said, sharply. Elizabeth in the dark hall smiled, and turned back to the living-room. Mrs. Reilly thought of the bus. In the dimness then she saw the chair, the cane, the silky pink dressing-gown. She began to move, and heard the gentle rain.